Quantum Holographic Criminology

Quantum Holographic Criminology

Paradigm Shift in Criminology, Law, and Transformative Justice

Dragan Milovanovic

Justice Studies Department
Northeastern Illinois University

Carolina Academic Press
Durham, North Carolina

Library of Congress Cataloging-in-Publication Data

Milovanovic, Dragan, 1948-
Quantum holographic criminology : paradigm shift in criminology, law, and transformative justice / Dragan Milovanovic.
pages cm
Includes bibliographical references and index.
ISBN 978-1-61163-481-5 (alk. paper)
1. Criminology--Methodology. 2. Holography. I. Title.

HV6025.M54 2014
364.01--dc23

2014008427

Carolina Academic Press
700 Kent Street
Durham, North Carolina 27701
Telephone (919) 489-7486
Fax (919) 493-5668
www.cap-press.com

Printed in the United States of America

Contents

Acknowledgments

This book has been inspired by major developments in quantum and holography theory. It is a challenge to Newtonian-based ontologies. We have not met the challenge in the social sciences. I have great pleasure to acknowledge the few true pioneers in meeting this challenge: Raymond Bradley, Stephen Robbins, and Alexander Wendt. They have initiated a new era of thought and have been inspirational for writing this book. Challenging the dominant disciplines in their innovative theorizing, they, often with singular conviction, have taken the initial steps in truly engaging quantum and holography theory in the social sciences. My work here integrates, builds upon, and in many cases, reconceptualizes many of their thoughts. Thanks to David Polizzi, editor of the *Journal of Theoretical and Philosophical Criminology*, who saw the merits of my initiative. I owe many thanks to Joyce Kelly, who was always there in support and encouragement for the challenging task before me, and with whom I often shared the "aha!" moments of excitement. Beth Hall, editor at Carolina Academic Press, has been unwavering in her support, encouragement, patience, and excitement for my project. She is a truly exceptional professional editor with whom to work. Special thanks, too, for Grace Pledger, production editor at Carolina Academic Press, for her impeccably prompt, patient, and helpful responses. Any errors, however, are mine.

There was a time that Newton's great discoveries were promising to produce a Laplacian deterministic universe, which, theoretically, could lead to discoveries of all laws of nature (Albert Einstein's "God Does Not Play Dice"), and in particular, all causes of crime. The latter ideal has been uncritically incorporated in contemporary dominant forms of empirical and positivistic research. We need to reconsider in light of its failures and the challenging new emergent sciences. Thus, this is also an anticipatory acknowledgment of those to follow, brave and bold enough to rethink ontology and epistemology and to recognize the unique becomings, "clouds of possibilities," that we are, and the more just society that is possible.

Quantum Holographic Criminology

Chapter 1

Introduction

A legacy of adhering to the principles of Isaac Newton has led us to a standstill in the social sciences, particularly in criminology, law and social justice. A new physics has emerged with many scholars in the social sciences merely curious onlookers. This is the world of quantum and holography theory. Principles of quantum theory were established in the 1920s, and to this day, none of the main principles have been found to be suspect. It is indeed a "weird science" that fundamentally questions our much beloved core ontological assumptions. Holography theory, having its developmental history traced to the mid-1940s, 1980s, and 1990s, has offered even stranger notions of the world we thought we knew. Tremendous understandings have been engendered due to these two insights, but yet, the social sciences rigidly adhere to the fundamental principles of Isaac Newton, established in his *Principia* in 1687. Yes, Newton's principles certainly ushered in a paradigm shift in the understandings at that time. And these fundamental principles were, over the years, incorporated into the social sciences as bedrock assumptions. They explain everything at the observable macroscopic level, from driving a car, to taking a walk, to launching a rocket to the moon and beyond. This calculus seemed over time to be quite adequate in explaining movement in time and space. In the social sciences, we witness quantitative methodology that is fundamentally rooted in the ontology suggested by Newton's calculus with a little help of classical probability theory. Qualitative methods, too, assume these fundamental postulates when trying to understand the wherewithal of various phenomena. Empiricists, and so too social scientists generally, have embraced the full logic of the determinism inherent in Newton's physics in constructing a positivist framework of cause and effect. It underlies our constructions of social reality. But things have changed in the physics and cosmology scholarly communities.

Albert Einstein, of course, was to offer his theory of relativity in 1905 and 1915, which challenged many of the core assumptions of Newtonian physics. But yet, Einstein insisted on a deterministic universe, best expressed in his quip, "God does not play dice." Certainly, some versions of relativity theory have found their way into social science thought. But quantum theory was to challenge even this fundamental paradigm shift. In their studies of the subatomic world they found that a new set of understandings were needed to fully appreciate what is at work. Many of these findings at the sub-atomic level to this day are not fully understood, to which many adherents reply that it is indeed a weird science. Nevertheless, its principles continue to receive universal support. But, you say, that deals with the subatomic level. What possible usefulness does quantum theory have with the more macro-level of observables? What indeed does it have to do with violent crime, judges deciding a case in law, warring gangs, and even notions of social justice? We shall see that our constructions of images, which are the basis for making sense of and acting in the world, are based on quantum and holography principles. In understanding our social construction of reality we necessarily must look at how what we see ultimately as "objects" are constituted out of vibratory entities both emitting and receiving electromagnetic radiation. These are the carriers of encoded information.

Quantum theory has shown that everything can be understood as a quantum wave function, a probability function. It stipulates that all can be conceptualized as a moving "cloud of possibilities." The central question becomes how does this cloud "collapse" in instantiating what we perceive as "reality"? In other words, one instance from the many possibilities. Erwin Schrödinger in the 1920s developed the equations to explain this movement and possible materialization. Quantum probability theory rather than classic probability theory was necessary to understand these materializations. In response to "God does not play dice," physicist Bohr scolded Einstein, "Don't tell God what to do." The quantum wave function was a mathematical description explaining the cloud of possibilities in movement. But it was de Broglie in 1924 who suggested that it is not just a mathematical description, but also has a physicality. Those that follow, beginning with Bohm (1983), were to show that indeed the wave function incorporates embedded holographic information.

Consider an example of the working of the wave function. "Interrupters," often former gang members, who now work with gangs in trying to interrupt their otherwise constructions of social reality that often justify violent retaliation, actually are the agents who collapse the wave function differently. From the cloud of possibilities an alternative construction of social reality is constituted, which suggests that a gang member was not "dissed" and that there was another explanation. So an otherwise escalating dynamic leading to yet another shooting on the West Side of Chicago has been avoided.

Let's look at the second major development since relativity's challenge of the Newtonian universe. Holography theory, well accepted in physics and cosmology, has yet to be taken seriously in criminology, law, and social justice. There is some insightful work being done in the study of word codification in the brain by psychologists, and some work following Karl Pribram's findings of the brain operating by holography principles. In the social sciences there is some work being done which has yet to take off. Explorers Alexander Wendt (2006, 2010) in political science, studying international relations, and Raymond Bradley (1998, 2010, 2011) in sociology, studying group solidarity, intuition, and terrorist cells, are two of the very few innovative theorists following the principles of quantum and holography theory.

But what is holography theory and why is it relevant? Many of us are familiar with Princess Leia's appearance as a projection in the early *Star Wars* episodes. Many of us, too, marvel at credit card logo or pictures we often find in specialty stores that make use of holography. We are intrigued at the way a three-dimensional picture emerges. We shall show that the very fundamental principles at work here are also at work in how social reality is constructed. Now this might seem initially a jump in logic. But let us pause. Consider the DVD you play on your disc player. We look at the disc and we see a flat, shiny surface. Nothing unusual. We look even closer by appropriate instruments and we see nothing that suggests the eventual appearance of visual and auditory information. But yet, we insert the disc and listen and observe a Bruce Springstein concert that took place in Barcelona, Spain. It's almost as if we were there, especially if we are playing a high definition or blue ray mode. Now this is not holography, but does suggest how a flat, insignificant-looking surface can generate such a rich text. A hologram is created by an interference pattern between two wave patterns that intersect in a particular way. This interference pattern can be encoded on some recording surface. Looking at this two-dimensional surface, we see scribbly lines that suggest little to us. But yet, when we shine a beam on any portion of our recording surface the entire initial image re-appears in three-dimensional form (e.g., Princess Leia appears). What is remarkable is that the image is encoded all along the recording surface. Illuminate any part, and the whole image appears. Could it be, by the same logic, that the "out there" is much like this recording surface (interference patterns) encoded with information? And we, as agents, access and transform this information into a three-dimensional everyday reality? This is startling, but yet, it is at the basis of holography theory.

Well, all very interesting, but what has it to do with criminology, law and social justice? We will show that holography theory establishes how information is encoded, stored, transmitted, and interpreted. Those doing criminology, law, and justice would gain a tremendous understanding of their undertakings if they were to entertain the possibility that this work is fundamentally about holography theory. We draw from the incredibly insightful work by Henri Bergson who in 1896 wrote *Matter and Memory*. This, we will argue, anticipated many of the principles of holography theory. These possible connections were explored by Milic Capek (1971), who connected Bergson's work to quantum theory, and Stephen Robbins (2006, 2010, 2012), in quantum holography theory. Deleuze and Guattari in the 1970s and 1980s were to draw heavily from Bergson to provide a further suggestive approach that lends itself to a holographic integration. Much of Deleuze and Guattari's work, we will argue, is implicitly about quantum and holography theory.

Abner Shimony (1965) was one of the first to demonstrate a connection between early quantum theory and the social sciences in the work of Alfred Whitehead (1925, 1929). Generally, physicists and cosmologists that reference consciousness embrace Whitehead. In our view, however, it is Henri Bergson's work that implicitly offers more perceptive commentary on both the quantum and holographic nature of the cosmos.

In understanding agency, or the wherewithal of a person's consciousness, holography theory offers some alternative notions as to how information is encoded, stored, transmitted, and interpreted. In addressing the question of the missing agent in much theorizing in criminology, law and social justice, we will offer a de-oedipalized, quantized version of Jacques Lacan's (a revisionist Freudian psychoanalyst) model of an inter- and intra-subjectively constituted subject. We will call our modified version schema QD. It is schema QD that when instantiated produces an emitting signature wave that interacts with the "out there" in constituting it where it is, "out there." This follows Bergson's notion that consciousness is not *of* something, but *in* something. We are left with the conclusion that everything with which we interact remains with a trace of our signature wave. We are all part of the universe. This is inescapable. We are intricately interconnected with the cosmos. If this is so, dear reader, you can see how thinking in conventional criminology, law and social justice is plainly disconnected and disembodied, leaving the nature of being human a mere shell of what it is and what it can possibly do and become.

Accordingly, this book will lay out the core principles of quantum and holography theory. It will compare these with the ubiquitous Newtonian-driven ontology. We suggest that the latter is the basis of a classic materialist paradigm. We will then further develop quantum and holography principles and integrate them with, primarily, the work of Bergson, Deleuze and Guattari. We will develop a quantum and holography inspired notion of agency, schema QD. Our last two chapters will be concerned with applying this theoretical material to law, criminology and the development of a transformative justice. This will be for expository and pedagogical purposes. We make no claim for developing complete pictures of theories of crime, nor for a blueprint for an alternative form of law and transformative justice. Rather, ours suggests a reorientation and some novel guidelines.

Ours is a journey that has just begun. Fortunately there have been inspirational explorers: Milic Capek, Abner Shimony, David Bohm, Karl Pribram, Raymond Bradley, Stephen Robbins, Alexander Wendt, to name some of the more prominent, and other pioneers in quantum cognition theory who describe the nature of the human lexicon structure. Other pioneers in developing a quantum holography-based understanding of the encoding, storage, transmission, and reconstruction of information include Peter Marcer, Walter Schempp, Ervin Laszlo, and Edgar Mitchell. Let's rise to the challenge. Our criminal justice system has been recognized as a failure for some time. Any system that generates a two-thirds recidivism rate (ex-offenders re-arrested), imprisons over two million of its citizens, and has a disproportionate impact upon African-Americans is plainly a failure. But it continues, given full thought to the idea that perhaps in its failure it is a success for dominant groups. Let's see how the new principles of quantum and holography theory will begin to release us from our shackles.

Chapter 2 develops the salient differences between the classical-materialist paradigm and the process-informational paradigm. The former is rooted more in Newtonian physics, the latter more in quantum and holography theory. We first identify key figures within each, followed by commentary on their importance to each paradigm. We then compare the two paradigms along selected distinguishing dimensions: the nature of objects, space, time, cause, information, language, and processor of information. Each comparison is further organized in terms of: key concepts, commentary, and applications to criminology. This is not meant to be exhaustive but an introduction to the profound differences existing between the two paradigms. It contributes to not only situating contemporary theorizing in criminology in terms of restrictive core assumptions, but provides novel conceptualizations with which we may engage more critical analysis.

Chapter 3 develops the notion of the holographic principle, primacy of information, communication based on phase conjugation, and how to conceptualize entities in terms of assemblage theory. There are at least four prominent holographic principles being currently employed. Karl Pribram's (1971, 1991) model suggests that memory storage is in the brain; Bohm's (1980) model suggests that it resides in an "implicate order," a holofield that can be actualized as an explicate order; Mitchell (2011), Laszlo (2007), Marcer and Schempp (1997, 1998, 1999) emphasize that memory resides in the "in-formation" field, the quantum space within which all are situated and is constantly changing; and the fourth, derived from studying the black hole and information retention, argues information is stored on boundaries rather than the internal volume ('t Hooft, 1993; Susskind, 1995). Gabor's (1946) early statement of how information, as geometrically encoded patterns of energy, follows the Heisenberg uncertainty principle, shows that information is inherently in-process in usage. We then move to the notion of "phase conjugation" in indicating how information channels are established. All entities are vibratory in nature, and any interaction between and among entities concerns in-phase or out-of-phase resonance. Where in-phase, we have commonalities established and possibilities of meaningful communication. We review two exemplary studies of a quantum holographic approach in the social sciences, one in international relations indicating how members of a collective instantiate the state, the other in sociology dealing with the nature of solidarity. We then move onto a novel way of conceptualizing all entities. These are assemblages developed in the work by Deleuze and Guattari (1987). We shall summarize the key components and different forms assemblages may take. We find that each assemblage generates a logic, an "abstract machine," that has effects on other assemblages. These "lines of flight" we will equate with vibratory energy encoding holographic information. These are "affordances" (Gibson, 1979), or restricted possibility waves. We conclude the chapter by providing examples of quantum holography at work in the social sciences and in criminology in particular. The quantum holographic perspective on the various approaches selected is meant for expository purposes, not a definitive statement of a finished product. These provide the reader further thought on the subject, and hopefully the concepts that can produce a corrective to the entrenched Newtonian-based core assumptions with which the social sciences are embraced.

Chapter 4 focuses on the nature of agency. The social sciences and criminology especially are vacuous as to a bona fide concept of agency. Positivism *or* rational choice theories are the borders for discussion. We suggest an alternative inter- and intra-subjective construction of the agent, which we refer to as schema QD. This is a de-oedipalized, quantized adaptation of Jacques Lacan's (1977) schema R. Before developing this notion, we briefly look at two prominent models of a quantum consciousness, one offered by Penrose and Hameroff (2011) in physics and neurobiology indicating the collapse of the wave function is brought about objectively when some threshold value of intensity collapses producing conscious states; the other by Henry Stapp (2007), built on the more pragmatic Copenhagen version of the collapse of the wave function and on prominent quantum physicist Newmann's (1932) ideas of an "abstract ego," which suggests that the agent poses questions to nature, which responds with answers, from which a particular state of the wave function results. We then move to developing schema QD. We provide the various interacting components and suggest how a speaking subject, an "I," is instantiated. We show that this instantiated "I" is represented by a distinct signature wave that interacts with others. Phase conjugation and the insights of Henri Bergson are integrated in showing how perception-images are created. We show how information is accessed from the in-formation field and momentarily stabilized in perception-images, or "logons," the basis of social action. We then borrow from topology theory to map schema QD on a "cross-cap" which is suggestive as to the types of "cuts" of the subject that may be instantiated. Each cut of the subject is a collapse of the wave function. We draw from a small group of quantum cognition theorists who describe the nature of lexicon storage by way of quantum and/or holography theory. We close the chapter in indicating how political economy is often an axiomatic system "painted" on its boundaries, from which affordances are mobilized, and which provide the basis of loci of subjectification (constructions of circumscribed identities). In more passive modes of reception, reification results; more active attention, on the other hand, becomes the basis of "creative evolution."

Chapter 5 turns to applications in criminology and law. We apply the concepts to which we have been introduced to some key, foundational theories of crime in indicating a better understanding of the physicality of the theories suggested. Cultural criminology, perhaps currently the best show in town in taking a holistic approach in critical criminology, will be our focus in the criminology section. We look at political economy, questions of agency, forms of Othering, cultural/subcultural practices, media practices, the allure of edgework and emotionality. Two key works are introduced, one on the relation of gangster rap music to violence, the other on terrorists and terrorist cells. We are especially mindful of a constitutive framework that suggests co-production of social reality. Concepts developed in Chapters 2, 3 and 4 are now applied in demonstrating their worth in doing critical criminology. We then turn to law. We focus on a recent Bergson/Deleuze driven theory of judges' decision-making (Lefebvre, 2008). We introduce schema QD in showing that it provides more insightful ways of seeing how reality construction and decision-making takes place. From the various possibilities that the quantum wave function represents, a judge collapses the wave function by her/his attention to one possibility, making it a case with legal standing. Once established with legal standing and subsumed under some principle in law, deductive logic and syllogistic reasoning leads to linear derived truths in law, a decision in law. We show how phase conjugation and accessing the legal in-formation field takes place. We explain how the judge, as a speaking subject, constructs narratives from a limited range of a legal lexicon. We, however, introduce research showing how nonlinear decision-making can take place, producing changes in law (away from linear constructions based on *stare decisis*). This chapter, once again, is suggestive for the type of research we could undertake. It is not exhaustive, nor does it try and develop a self-standing theory. It points to work that is a necessary break from the imprisonment we find ourselves in.

Chapter 6 applies quantum holography conceptualizations to transformative justice. We recognize the limitations of restorative justice, especially in its internal constraints due to dependence on the core assumptions of Newtonian physics. It also, as many commentators have argued, has the potential for "widening the net," disciplining its participants in the language of restoration, and being redefined within the constraints of "criminal justice." We first turn to the notion of "molecular assemblages." We examine its ubiquitous nature in the social field (socius). We show how phase conjugation tends to reify dominant orders. We then examine models which may provide a phase space in which molecular assemblages may flourish. These instantiations are more along the lines of "dissipative structures" that dynamic systems theory (chaos) suggest, both providing momentary order, but always in process of dissipating and rebuilding, or in Deleuze and Guattari's language, deterritorialization and reterritorialization. We seek a model of permanent change, in which both relative stability and becoming-other is ubiquitous. We re-examine a recent Deleuzian model of the socius with Holland's (2011) "nomad citizenship" for possible directions toward a more just society. Applying schema QD we will show how homeostatic-driven discursive subject positions, or spaces within which an "I" can take up residence to speak, are more restrictive in molar assemblages, but in continual variation in molecular assemblages. We then examine the differences between truth constructions: the "man of truth" discovers truth, s/he is the purveyor of correct understandings and acts upon this understanding; the power of the false, on the other hand, suggests truth is not discovered, but is an ongoing creation. We show how different forms of assemblages can be the basis of different "cuts" of blocs of spacetime, producing particular notions of social reality. Spacetime constrained "logons," as encoded information, have more dynamic states in molecular form of assemblages and sociuses, and conversely in molar assemblages.

In our final section, examined are possible evaluative schemas. This addresses the paucity of normative theory in much of progressive literature. We provide an initial approximation of a distributive and a retributive ethical justice principle. These are suggestive guidelines by which we could evaluate existing active and reactive forces in a socius. We are inspired by the work of Spinoza (1985) who recognizes that we still don't know what a human being is capable of doing and becoming, and his notion of cultivating joyful passions. We look at how evaluation takes place during phase conjugation and the dynamics of schema QD. An alternative society-in-process is one in which we find, with Deleuze (1986, 1989), a "people yet to come," a continuous becoming-other. Information in the emerging molecular driven socius will witness instantiations of the wave function that are expansive, representing genuine possibilities that the human represents. Phase conjugation will provide not linear reproductions of what has been holographically encoded in the information field, but, following Gabor "spacetime constrained logons," will always have an indeterminate quality; hence, the agent is never a determined being. The very last section provides some commentary on a key note address driven by quantum holographic principles that was delivered by this author to a conference "Crime and Social Policy in the Caribbean." It suggested directions away from a Newtonian-based criminal justice toward a transformative justice model.

Let's begin our journey. Let us suspend prevailing Newtonian-driven ontological assumptions and entertain the novel concepts offered by the new sciences of quantum and holography theory.

Chapter 2

Classical-Materialist Paradigm v. Process-Informational Paradigm

Introduction

Social sciences are at a standstill. They remain embedded in frameworks that resist critical reflection.[1] Dominant contemporary criminology and a good part of critical criminology, too, remain in frameworks that deny recent developments in the physics community. More specifically, Newtonian physics remains a bedrock core of assumptions resisting the findings of quantum and holographic theory. In this chapter we want to encourage rethinking ontological assumptions. We therefore outline two paradigms. These should not be seen as dichotomous, for there are overlap areas. Nevertheless, for pedagogical purposes, conceptualizing key parameters of each orientation provides for critical analysis of core assumptions. This chapter has been inspired by the work of Whitehead (1925, 1929) who compared a Newtonian-based "organic cosmos" to a "quantum cosmos," and equally by Deleuze's (1986, 1989) two-volume opus on cinema which suggests a more Newtonian "movement image" compared to a more quantum and holography process-oriented "time image." Deleuze has also argued that a particular system of signs follows these two notions of space and time: the former as the "organic regime," the latter, the "crystalline image." We conceptualize these "spaces" as noospheres populated by noosigns.

Criminological literature has offered other comparisons, devoid of a discussion of ontological elements founded by Newtonian compared to quantum theory. Some early and contemporary discussions in critical criminology include: consensus v. conflict theory; consensus v. Marxist; consensus, pluralist, radical; consensus criminology v. critical criminology, etc.

1. Some research has recently surfaced, however, in psychology disciplines, particularly on storage and retrieval of memories via either quantum or holography dynamics.

We will develop two paradigms, the classical materialist and the process informational, along eight dimensions of comparisons.[2] This is not to say there are not other dimensions by which we could compare. We could conceivably even collapse some of these dimensions. This is a first approximation. Thus our comparative ontological dimensions include: prominent founding theorists; nature of objectivity, time, space, causality; construction, storage and transmission of information; nature of language; and how we conceptualize the processor of information. Each comparison will be divided into three: key concepts, commentary, and applications to criminology (critical commentary), the latter more suggestive than exhaustive.

Classical-Materialist Paradigm, Process-Informational Paradigm: Perspectives Compared

1. Prominent Founding Theorists

a. Classical-Materialist Paradigm (Key Theorists)

Key theorists. Newton, Descartes, Euclid, Laplace.

Commentary. The classical-materialist paradigm is centered on the work of Isaac Newton (1643–1727). His *Principia* (1687) provided key principles for conceptualizing particles and their movement in space and time. These were the laws of motion. Along with Gottfried Leibniz, he formalized calculus. These offered an ontology for a rational approach in the examination of physical properties and dynamics of the cosmos. René Descartes (1596–1650) provided additional components. He systematized an analytic geometry and a Cartesian coordinate system. Euclid (born 300 B.C.), from ancient Greece, often seen as the "father of geometry," in his influential *Elements* provided the main principles for what is now referred to as Euclidean geometry. From a small set of axioms he drew, by deduction, various principles of geometry. Together Newton, Descartes, and Euclid provided the key elements of a classical-materialist paradigm that were also incorporated in the social sciences as foundational principles. Pierre Simon Laplace (1749–1827), often referred to as the "French Newton," was perhaps the exemplary voice of the necessary logic

2. References for and discussion of concepts in the following sections concerning "key points" will be provided throughout this chapter and book. See also our index for locations of various discussions.

of this development, a clockwork-like universe, where ultimately all can be predicted. This was systematized in his *A Philosophical Essay on Probabilities* (1814). He posited an intellect, "Laplace's demon," capable in theory of ultimate prediction of all activity.

> We may regard the present state of the universe as the effect of its past and the cause of its future. An intellect which at a certain moment would know all forces that set nature in motion, and all positions of all items of which nature is composed, if this intellect were also vast enough to submit these data to analysis, it would embrace in a single formula the movements of the greatest bodies of the universe and those of the tiniest atom; for such an intellect nothing would be uncertain and the future just like the past would be present before its eyes.

b. Process-Informational Paradigm (Key Theorists)

Key theorists. Early quantum mechanics: Plank, Einstein, Bohr; post-1924: de Broglie, Bohr, Born, Schrödinger, Heisenberg, Pauli, Diric, von Newmann; quantum holography: Bohm, Laszlo, Gabor, 't Hooft, Susskind, Bekenstein, Maldacena, Witten, Bousso; compatible philosophers: Leibniz, Spinoza, Bergson, Whitehead, Deleuze, Guattari.

Commentary. Quantum and holography theory are core contributions in the process-informational paradigm. These formulations directly challenged the key concepts of the classical-materialist paradigm. Plank, in 1900, conceptualized light as made up of quanta or packets. Albert Einstein, in his general and special theory of relativity (1905, 1915) was to overturn many of the core concepts of Newtonian physics, showing that time and space are connected (i.e., space-time), that energy and matter are interchangeable, and that the cosmos is inherently relative, not absolute. His (1905) further work on the photoelectric cell was to provide important foundational material for the coming quantum mechanics. Light was seen as composed of photons that exist in discrete packages. Ironically, Einstein was subsequently opposed to the end to the "weird" science of quantum mechanics. Bohr was to provide a model of the atom with its circulating electrons. He showed that matter was not solid and that an atom was comprised of an incredible large amount of space. But it was from 1924 onward that a more mature quantum mechanics was to develop with many of its core principles established between 1924–1928. De Broglie, Bohr, Schrödinger, Heisenberg, Pauli, and Diric were to provide key provisions. John Von Newmann (1903–1957) provided much of the formal mathematics for quantum theory that evolved in the 1920s in his book *The Mathematical Foundations of Quan-*

tum Mechanics (1932). This, too, became a center piece for one version of quantum mechanics, the Copenhagen interpretation.

Holographic theory has developed along four relatively independent lines. The first was by Denis Gabor in 1946. Gabor was working to improve transmission of information and was to posit how quantum mechanics and holography work together. He applied Heisenberg's uncertainty principle (one cannot measure with exact precision *both* momentum *and* time of something in motion). He demonstrated that the daily, ongoing production of information was attributable to both spectra (interference patterns established by the intersections of electromagnetic radiation) and spacetime constraints that produce "logons," quanta of information. He was to influence Karl Pribram's (1971, 1991) work on the dynamics of the brain. The second direction came from Laszlo (1995, 1997, 2001, 2007) who suggested that we are immersed in a zero point field, an energy field that also is an in-formation field. The third direction was suggested by Bohm (1980; Bohm and Peat, 1987). He envisions the cosmos composed of a more hidden implicate order out of which an explicate order, or what we see as reality is instantiated. The fourth direction came from the study of black holes. Up until the early 1990s the dominant thinking was that anything that entered the black hole would disintegrate and be lost forever. David Susskind (1995) and 't Hooft (1993) working independently in the 1990s showed this was not the case. Following the second law of thermodynamics and entropy theory they indicated that all information of whatever falls in the black hole is preserved on its boundary, the event horizon. This is not what we would perceive as some kind of physical surface, but a mathematical imaginary surface. Bekenstein (2003) further refined this notion and computed the limits of information storage on the boundaries, called the Bekenstein bound. The new principle that emerged is called the "holographic principle." Subsequent research by Witten (1998), Bousso, and Maldacena showed conclusively that the holographic principle has mathematical support. Any region that has a boundary will have all its information stored on the boundary in two-dimensions. Holograms will populate the interior space (the "bulk") of this boundary region. We will develop these four approaches in greater depth in Chapter 3.

Several social philosophers' approaches have points of compatibility with quantum and holography theory. Shimony (1965) was the key theorist who showed that the social philosophy of Alfred Whitehead has compatibilities with quantum mechanics. This opened cross disciplinary integration and synthesis. Cosmologists and physicists who have ventured outside their discipline are most likely to invoke his name. Whitehead, mathematically trained, focused on the early form of quantum theory, even though he was a contemporary of developments of the more mature forms. Leibniz and Spinoza's work have also been invoked. Arguably, even more compatible with both quantum and holography theory is the work of Henri Bergson, particularly his *Matter and Memory* (1896) written a quarter of a century before quantum theory's and a half century before holography theory's formal development (see the clear connections developed by Robbins, 2000, 2006, 2010, 2012; Capek, 1961, 1971). More recently, the work of Deleuze and Guattari (1983, 1987), and particularly Deleuze's (1986, 1989) single-authored two-volume opus on cinema theory, is implicitly compatible with quantum holography, even though the latter terms are rarely invoked. Delanda (2006) and Massumi (1992) have been most noteworthy in making Deleuze's work more receptive to engagement with quantum holography (see also Robbins, 2000, 2006, 2010, 2012).

c. Applications to Criminology (Critical Commentary, Key Theorists)

Dominant criminology and a good part of critical criminology are in need of consultation with scientific developments since Newton and Descartes. Quantum mechanics developed primarily in the 1920s and holographic theory developed in the 1940s by Gabor and in a different form by 't Hooft, Susskind and Bekenstein in the 1990s offer radically new understandings of the cosmos. Whitehead's and Bergson's approaches share a clear ecological space. Theorizing by Deleuze and Guattari (1987) has provided suggestive innovations in philosophy, much of which is implicitly adaptive to quantum holographic principles. It is their work that offers the best hope for a fuller development of a process-informational paradigm and for its application in saving criminology from its malaise. Work applying quantum holography in criminology include Milovanovic (2011, 2012, 2013a, 2013b); in quantum theory and law (Kuttner, 2011; Kelsey, 2013; Mootz, 2013; Yager, 2013).

2. Objects

a. Classical-Materialist Paradigm (Objects)

Key concepts. Fixed in absolute time and place; enduring; "simple location"; essences; static; essentialism; objective; clear boundaries; particles precisely defined by location and momentum in x,y,z coordinates; set of instances.

Commentary. Objects are conceptualized as fixed entities, occupying specific space, objective in nature, with clear boundaries. They have essences that are measureable and are quantifiable. They are located in specific locations in time and space. The coordinate system for their location is Euclidean space composed of x, y, z coordinates within which all movement could be located and precisely measured. Their movements can be precisely traced from one moment to another. They are seen as instances that can be precisely measured. They stand alone as distinct from other objects. This is a science of exactness, objective reality, a realist ontology. Things simply are; they exist in themselves, or not. Science attempts to eliminate all biases and contaminants to measurement until the object itself is clearly revealed and specified, that is, identified as "data." Confounds and threats to internal and external validity are the sources of inaccurate data, and theoretically, can be eliminated.

b. Process-Informational Paradigm (Objects)

Key concepts. Process; events/occasions; vibratory; wave function; interconnectedness; wavicle; wave collapse; uncertainty; measurement problem; emergent; delocalized; entangled; quantum coherence; dynamic; assemblage; frequency, wave emitters/absorbers; dissipative structures; contingent; "smears/clouds of possibilities" (Ψ); actual/possible; actual/virtual; actual entities/occasions.

Commentary. Objects have no clear essences, nor can they be precisely measured. This is a science of indeterminacy and probabilities, not exactness. Objects are inherently vibratory, concentrations of energy, that do not occupy distinct space, nor can they be precisely located in time. Things don't exist until they are measured or observed. This is called the "measurement problem." Heisenberg's uncertainty principle shows one cannot measure location and momentum at the same time; measure one precisely, the other will resist precise measurement. Each entity absorbs and emits energy from its surrounding producing distinct effects on its character. Each is interconnected with all other entities and because of mutual perturbation, is more a dissipative structure in "essence"; an ever becoming. Each can be quantum entangled with other entities and at any moment may remain in a state of quantum coherence where contradictory states exist side by side. Each is characterized by a unique frequency emission; it is itself a "wavicle" identified as a plurality of possibilities, or sometime called a "smear" or "cloud" of possibilities, designated as Ψ. Schrödinger, in the late 1920s, formulated the notion of quantum waves to characterize each entity. Each entity has its own unique quantum wave or quantum signature; larger entities can be composites of several wave functions. An entity could be an object, atom, proton, electron, etc. It can be a person, institution, society, etc. Words, thoughts, ideas, as well as an encounter can also be represented by a quantum wave function. A "cloud of possibility" is thus virtual; it is instantiated as a real only with the collapse of the wave function.

Schrödinger's classic cat-in-the-box thought experiment and Thomas Young's (1804) double-slit experiment (see demonstration by Matthew and Davidson, 2006) and its many confirmations in the literature were pivotal in this new conceptualization of how objects can be perceived. In the thought experiment of a cat in the box, a cyanide tablet is set to go off with the decay of a radioactive substance. Open the box and one will find the cat either dead or alive. While within the box, the cat can be both alive *and* dead. The act of observation (i.e., opening the box to view its contents), or measurement, produces an effect: the cat being either alive or dead.

With the "double-slit," or "two-slit experiment," a photon or electron beam is sent through two parallel slits onto a recording surface.[3] They pass through each of the slits and create an interference pattern on the other side that is then recorded on a surface. Interference is created when two high points of the two waves match. These interference patterns appear as multiple white vertical bands. Where they do not, that is the low point of one wave and the high point of the other meet, they cancel each other out: they produce dark vertical bands on the recording surface. The pattern that emerges from the beams clearly shows a wave pattern on the recording surface. If it was more "particle" like, there should not be several light vertical bands separated by dark vertical bands. Here we would have expected that two photon/electrons that pass through the respective slits should produce only one white vertical band for each of them. But this is not the case. In other words, if we were to have only one slit, then on the recording surface we would only find one light vertical band that suggests the particle like nature of photons. It is when we introduce two parallel slits that we experience wave interference patterns on the recording surface, thus suggesting the wave-like nature of the photon. Any experimental attempt to find which of the two slits the photon has passed through (to determine if it is a particle) collapses the interference pattern (the wave aspect), producing a particle. In other words, if an experimenter was to place a recording device at the slits to determine in what state the electrons exist at that point, they will find particle like patterns on the recording surface (i.e., two white vertical bands, each corresponding to one or the other slit). Thus the conclusion: the act of measurement or observation collapses the wave function to a particle-like entity. More recent versions of the double slit experiment include "quantum erasure" (Scully and Druh, 1982; Kwiat *et al.*, 1992; Walborn *et al.*, 2002; Cramer, 1998; Aharonov and Zubairy, 2005) and "delayed choice" (Wheeler, 1983; Alley *et al.*, 1987; Jacques *et al.*, 2007).[4] Experiments in quantum erasure show con-

---

3. Several excellent videos are available on the internet: see http://www.youtube.com/watch?v=LW6Mq352f0E; http://www.falstad.com/ripple/ex-2slit.html; http://homepage.univie.ac.at/Franz.Embacher/KinderUni2005/waves.gif; for a more animated, but yet clear demonstration, see: http://www.wimp.com/drquantum/.

4. See, for example, Campbell's excerpted lecture, at 1.04.41 to 1:13.45, http://www.youtube.com/watch?v=vU6yCD_sEvU.

clusively that if one eliminates the initial recording device as to which slit a photon may have passed through the interference pattern is once more recreated. It is the presence or absence of the measurement/observation itself that either created the interference pattern or not! At the macro scale, however, all quantum characteristics collapse to classic physics, that is, Newtonian characteristics. Thus the double-slit experiments along with Schrödinger's thought experiment of the cat-in-the-box are fundamental in understanding many more abstract or mathematically derived formulations of quantum theory.

Whitehead (1929) has referred to instantiated entities as "actual occasions" to suggest their emerging and dissipative characteristics. Smolin (2001: 65) has said "elementary particles are not static objects just sitting there, but processes carrying little bits of information between events with which they interact, giving rise to new processes." Bergson ([1896] 2002) has asked us to envision the cosmos as a dynamic soup of images, colliding, following distinct laws, out of which one image emerges that has the capacity to delay action. This incorporates the "zone of indetermination" that is unique to each "living image." Deleuze (1986, 1989), building on Bergson, has argued that matter=energy=light (recall Einstein's classic equivalence of energy and matter in $E=mc^2$). Deleuze and Guattari (1987) have also argued all entities are unique assemblages constituted by forces of stability and instability as well as by forms of expression in relation to material perturbations (we shall return to an elaboration of "assemblages" in the next chapter).

c. Applications to Criminology (Critical Commentary, Objects)

Classical materialist-based criminology assumes a static universe within which all "objects" can be, in principle, given precise quantification. A premium is placed on ever more precise operationalizations of variables. Variables are reduced to isolated entities that are stand-alones. A particular snapshot, a bloc of spacetime, isolates objects in their relational patterns which can then be mapped onto diagrams such as in regression and path analysis, with each variable contributing to "variance explained." In law, we witness legislative acts as well as case law where complex concepts are reduced to particular dimensions that are then defined as the appropriate understandings (see, for example, *Black's Law Dictionary*). Process informational-based criminology, on the other, would privilege the many facets that characterize an entity knowing full well that any pragmatically driven operationalization is but one privileged (real) amongst many possible (virtual). Entities would need to be "captured" not in a snapshot, but in a dynamic, ever changing flow. Jacques Derrida (1998) has suggested that concepts exist in opposition, where one side is privileged at the cost of the other which nevertheless remains an insisting force. This is also along the lines of Karl Marx's view of the dialectical nature of an entity, a composite of opposites. It calls for an operationalization of a variable that acknowledges the underlying tension existing within any entity. Both are beginning expressions of a process-informational driven approach, but have not been integrated into contemporary dominant nor critical criminology.

3. Space/Fields

a. Classical-Materialist Paradigm (Space/Fields)

Key concepts. Absolute; flat; Euclidean; Cartesian, 3-D, x,y,z coordinates; empty; homogeneous; traditional four—electromagnetic, strong and weak nuclear forces, gravitational; hodological; void; striated.

Commentary. Space is characterized in terms of functions for describing positions and trajectories of particles (objects). Space itself is conceptualized as predominantly "empty," devoid of substance. For instance, adherents point out, look at an atom. If the nucleus was a grain of sand the electrons would be orbiting the outside of a football stadium with "nothing" in between. It is homogeneous in form, similar everywhere. It is three-dimensional having length, width, height. A point has zero dimensions; a line, one; a plane, two; a sphere, three. Euclidean geometry situates all entities in an x,y,z coordinate system. Space is also seen as a field. Lewin (2007, 2010), for example, suggested a "hodological" field, a "life space," in which vectors representing various forces converge and diverge producing particular effects. A researcher in this model can theoretically measure all vectors and thus predict the forces at work on a particular object, and forecast the potential results. It concerns objects, vector fields, trajectories and an ideal of predictability. Calculus would be the tool. The notion of a field has been given more precise measurement by the discovery of the four traditional forces recognized by classical physics: electromagnetic, strong nuclear forces, weak nuclear forces, and gravitational forces. Precise mathematics govern their respective behavior. They have also been combined in mathematical expressions, although gravity remains elusive in this integration. Thus space, fields and particles (objects) are the elements that provide the ontological basis of reality.

b. Process-Informational Paradigm (Space/Fields)

Key concepts. Curved; relational (Leibniz); network of relationships (Smolin); non-euclidean; fractal; multidimensional; topological; spacetime; quantum fluctuations; 4D/8D sub-space; holofield; ZPF; A-field; in-formation field; Higgs; network of nested holograms; torsion wavefield; implicate/explicate order; prehodological; holomonic; twister space; branes; light sheets; complex Minkowski space; Reimanian; r-space; multiverse; singularities; geographic; plane of immanence; ecological niches; conceptual spaces.

Commentary. Space is not flat or empty. It is teaming with energy fluctuations. Quantum fluctuations are ubiquitous in the cosmos. Various non-Newtonian conceptualizations exist (Jammer, 1954; Reichenback, 1958). Gottfried Leibniz (1991) saw it as a set of relations between objects. Smolin (2001), extending on Leibniz, conceptualizes it as a "network of relationships," so did Einstein with his notion of a spacetime affected by gravitational forces. Space doesn't pre-exist the kinds of relationships and linkages that exist. A point in space is the name we give at the intersections, and of specific features of these relationships (ibid., 22). But these sets of relationships and linkages are in constant flux. There are no fixed background spaces, and hence, particular features are always in process. A sentence (Smolin, 2001: 18), for example, is like space; it is a set of relationships amongst words; absent the words, there is no sentence, no space; space exists as a set of relationships. A similar example is provided by Sklar (1992: 20–23) following Leibniz (Jammer, 1954: 116–118) with relationships between members of a family; a family and its relationships cannot be separated. No relationships, no space. Think in terms of a college class and the set of relationships amongst students and with the professor. Visualization in Euclidean space is but an imposition due to our socialized proclivity to create geometric relations that aid in our pragmatic everyday actions (Reichenback, 1954: 37–58). Space, however, on another level is "full." The notion of a zero point field (ZPF) argues that every "point," or more appropriately, moment, in space is vibrant with energy. It is a vast reservoir of vibratory energy states. The notion of the Higgs field suggests there is a distinct milieu which interacts with all entities producing particular instantiations we see as the real.[5] The notion of the holofield (Bohm, 1983) suggests that the field is a storehouse of information encoding holograms. It is an "implicit order," a virtual space ("excitation patterns") from which the real ("manifestation") is instantiated as the "explicit order." It is an "in-formation" field (Laszlo, 1995, 2007) holographically encoding all information of the cosmos and is nonlocal. Nonlocal stands for the idea that something can have an effect on other things that are remote from it, where no specifically discernible factor can

5. We await further findings from the study of dark matter and dark energy to reappraise and/or revise our notions.

be identified at work. This field is prehodological, a set of potentialities. Within it, singularities exist which are moments where divergence is experienced.[6]

This space is also multidimensional, topological, and fractal. String theory, for example, as a possible conceptualization in developing "a theory of everything" exists in 11-dimensional space-time. At the "big bang" three dimensions of space and one of time were instantiated; the other dimensions were compactified in very small spaces, invisible to even the most sophisticated measuring instruments. The notion of the multiverse, as one extension, has it that multiple worlds exist, and at each singularity, divergent pathways will unfold each along its own time line. This space is often seen as fractal constituted by in-between dimensions. Thus among 0,1,2,3 of Euclidean geometric dimensions there are infinitely more dimensions, literally opening up space to multiplicity. This is the world of fractal geometry. Space is also topological, it is bent, sometimes bending back on itself, sometimes without recognizable boundaries, and takes on a plurality of geometric forms such as the Möbius band, Klein bottle, cross-cap, torus, etc.[7]

One attempt to connect an alternative space with consciousness, intentionality and nonlocality is by way of a "complex Minkowski space" that is not inconsistent with the mathematics of Newton, Einstein and Schrödinger (Rausher and Targ, 2001). This space is 8-dimensional: along with the three traditional (real) space dimensions and one of time there exist four corresponding imaginary dimensions. Complex numbers have a real and imaginary component. Imaginary numbers center on the issue of the square root of minus one. Consciousness, in this approach, has access to this higher dimensional space by way of intentionality and goal oriented behavior.[8]

In a previous section we alluded to the holographic principle which we will expand here and in the next chapter. There have been a number of versions: Karl Pribram (1971, 1991) building on Denis Gabor's (1946) classic work de-

---

6. Delanda (2002) shows how "symmetry breaking cascades" follow.

7. For an application in psychoanalysis, see Lacan, 1977; Granon-Lafont, 1985; Cochet, 1998; Ragland and Milovanovic, 2004. For Lacan, topology theory is not analogy or metaphor, it is the real.

8. Thus in this model, the complex Minkowski spacetime represents the locus of nonlocality and seemingly superluminal phenomena. Elitzur and Dolev (2006: 31) in their transactional model of the collapse of the wave function suggest that an extra dimension of hidden time may be needed, a "deeper time," to account for "time-outs" at which moment information is constructed and communicated. A complex Minkowski spacetime, in our view, is accommodative. Similarly, Batiz and Chauhan (2009) suggest that two extra dimensions might be needed to account for the movement of an emitting object wave and an attentional reference wave in accounting for the construction and transmission of information. Future work will investigate this further.

veloped a holographic model of brain space, indicating that the brain operates nonlocally; Bohm's (1983) work explains how space has two levels, the implicate order organized holographically, and the explicate order, the everyday instantiations that we see as reality; Laszlo (1995, 2007) has shown that not only the brain, but external space is holographically constituted; a number of theorists from physics and cosmology in studying the effects of substance falling into a black hole have concluded, consistent with the second law of thermodynamics and entropy theory, that all information of the otherwise disintegrating substance is preserved on the event horizon (Thorn, *et al.*, 1986; 't Hooft, 1993; Susskind, 1995, 2009; Maldacena, 1998, 1999, 2000; Bekenstein, 1981; Bousso, 1999, 2002). They suggest that all enclosed volumes (i.e., the "bulk" spacetime) with their populated three dimensional entities have their information encoded in two dimensions on its boundary surface. The surface acts like the recording surface of a hologram previously mentioned, the surface dynamics of which remains to be fully theorized, but nevertheless, the classical laws of Newtonian physics are but one outcome of its activity (see Verlinde, 2010). More recently, some (Bousso, 2002; Susskind, 2009) have argued that boundaries are light-like ("light sheets") similar to a "gravitational horizon," which embed holographic information. This idea has been incorporated in some versions of string theory to suggest that these surface boundaries are equivalent to "d-branes," vast imaginary sheets on which vibrating strings are attached, a basis for a multiverse.

In the social sciences, some potentially accommodative concepts of space include: geographic, ecological niches, contrast spaces, and a plane of immanence. Postmodern critical geographers (Lefebvre, 1974; Soja, 2011; see also Foucault, 1986) have offered a "spatialized ontology" which argues spatiality is a social product (Soja, 2011) a "set of relations between individuals and groups" (ibid., 120). Gibson (1979) has also offered the idea of "ecological niches" which are nuanced spatial locations with a characteristic vector field. The notion of a "contrast space" (Garfinkel, 1981; Salmon, 1984; Delanda, 2002: 129–134) suggests that human beings operate in relevancy structures from which they draw understanding and guides for action. These have also been conceived as "conceptual spaces" (Gardenfors, 2004). A classic example can be provided from Garfinkel (1981: 21): "when Willie Sutton was in prison, a priest who was trying to reform him asked him why he robbed banks. 'Well,' Sutton replied, 'that's where the money is.'" It wasn't an answer for the Priest concerning, from his conceptual space, why or why not he robs; but, from the transgressor's own contrast space, the key issue is the choice of targets. Deleuze and Guattari (1987) have developed the notion of a "plane of immanence" or "plane of consistency." It is an intensive field characterized by ongoing creation of differences, emergent heterogeneous assemblages, sometimes referred to as multiplicities, with various degrees of stability from ordered to orderly disordered such as in what chaos theory refers to as dissipative structures, and singularities (Delanda, 2002), or moments where symmetry-breaking cascades, period doubling (bifurcation diagrams of chaos theory), and what catastrophe theory envisions as abrupt "leaps," emerge. It is the virtual field from which the real is instantiated. In sum, these various notions of "space" from social philosophers have much to offer for a cross-disciplinary integration and understanding. They are central to a more process-informational ontology.

c. Applications to Criminology (Critical Commentary, Space/Fields)

Space and fields in contemporary dominant and much of critical criminology are framed in a categorical and passive manner whereby objects can be precisely measured. An offender's development can be conceptualized as a movement through a sequence of distinct spaces, a "life course," each with discoverable vector fields that contributed to her/his deviant outcome. These are seen as distinctively different from the law-abiding person's development. Ironically, even with this knowledge, when rehabilitation is attempted, it is ultimately directed to changing the person (object) rather than the contributing vector field(s). Critical criminology, on the other hand, focuses on changing the criminogenic vector field. With this goal, the vector field must be provided a metricized mapping, rendering it static. Conventional and often critical criminology are driven by a static, snapshot ontology. A process-informational infused approach would problematize space and seek its multidimensional and shifting vector nature. A geographic area or ecological niche would be seen as a repository of holographically stored information within a zero point field; alternatively, imbedded on the boundary surface of that space. Early work by Matza in *Delinquency and Drift* did suggest the "drift" nature of a juvenile's movement in and out of locations with an overabundance of neutralizations that contribute to delinquency. Recently developed cultural criminology is embracing this direction (Ferrell, *et al.*, 2008). Constitutive criminology (Henry and Milovanovic, 1996), arguing that the person is inextricably related to her/his environment, both constituting and being constituted by the environment within which s/he operates, has also been in this direction. Critical geography, awaits the integration of the suggestive work by Gibson (1979) and his notion of ecological niches, constitutive affordances, and ambient light as a carrier of ecologically constituted information. The process-informational ontology provides a more dynamic conceptualization of space and fields with which to conceptualize the directions suggested by Matza, emerging cultural criminology, critical geography, Gibson, and constitutive criminology.

4. Time

a. Classical-Materialist Paradigm (Time)

Key concepts. Absolute; discrete; metric; equal intervals; uniform; linear; forward, unfolding; objective; non-relational; clock time; irreversible; static/immobile cut; movement image.

Commentary. Time is understood as linear extending from a past to a present and unfolding into the future. It is uniform for everyone and everywhere. It is an absolute: static and objective. It unfolds everywhere with the precision of clockwork. Each moment itself is no different from any other. It is measurable and predictable. It is a metric time which includes uniformity, a unit of time, and simultaneity (Reichenback, 1958). It flows forward without any consideration of relations to actual events that transpire. It is functional to produce order, regularity, prediction, and certainty. Take any instance of objective time and you have an immobile cut, a snapshot of being. Thus we witness the affinity toward a snapshot social sciences, devoid of any notion of a perpetual becoming. This is most recognized in statistical analysis in their static operationalizations of variables which can be located in an absolute time. Time-series and cohort studies over time also assume an absolute time as a background space.

b. Process-Informational Paradigm (Time)

Key concepts. Relative; relational; topological; multiple time lines; timelessness; flow; spacetime; configuration time; backward time referral; duration/durée; event; time connected to change; mobile cuts; blocs of mobile spacetime; time-image.

Commentary. Time is not static, metric, homogenous, linear, discretely unfolding, and absolute. It is connected with changing configurations of relationships. There is no universal, absolute clock external to these changing configurations (Smolin, 2001). Space and time are interconnected: time can only be explained in relation to space which, following Leibniz, is composed of configurations of relationships (ibid., 24). There is no static background space such as in classical physics; all is movement and changing relationships. Einstein's great discovery, following the work of Leibniz's and Ernst Mach's critique of Newton's space and time, was that time is relative to the observer. Along with the traditional three dimensions of space, an additional one of time makes a 4-dimensional spacetime in which events take place. It is also topological (Reichenbach, 1958); it follows curvature of spacetime itself, sometimes looping, sometimes recursive. Gravity, Einstein showed, can bend spacetime. Time can also dilate as the speed of light is approached ("intrinsic time dilation"). Bergson (1999) was to go a step further in his critique of Einstein. Time is duration (durée); it contracts and dilates, even as we do not necessarily reach the speed of light. It differs in intensity. Each moment can be lived differently. Near death experiences and emergency situations, for example, show how time appears to slow down. A more radical position expressed by Julian Barbour (2001) in *The End of Time* is that there is no such thing as time; it is only a construct, a useful functional construct that is our collective, ongoing reification. Time is not linear and unfolding, but rather may move in forward and backward modes. The notion of "backward time referral" in quantum mechanics indicates how quantum information can be sent back in time to affect the present.

c. Applications to Criminology (Critical Commentary, Time)

Contemporary dominant criminology assumes time in classical-materialist terms. It remains entrenched in an untheorized background space and time within which human interactions take place. Time series, life course, cohort, and path analysis suggest a linear development with uniform absolute time. An event can be conceptualized in the three-dimensional Newtonian-Euclidean space with time added running uniformly from the past to the present to a future. A process-informational perspective, on the other hand, would look at the nonlinearity of time, retrocausal effects, and momentary intensities. Consider, in critique, *ex post facto* construction in law suggested by Left Realists (see summary by Milovanovic, 2003). A judge has an initial hunch, feel, or intuition in making a decision and then articulates an acceptable legalistic (semiotic) narrative to explain the decision rather than looking "objectively" at the "facts" of the case and then via deductive logic and syllogistic reasoning constructing a conclusion in law. Similar dynamics are exhibited in police behavior. Functionaries in the criminal justice system, too, know only too well how to dress up their activities in the allowable legal and bureaucratic language. On another plane, quantum theory has shown how we may move backward in time in having effects on present developments. We live in time and thus a snapshot criminology or law is insensitive to the ever becoming of the human being. Rather, a mobile "cut" of spacetime is suggested. Deleuze, for example, criticizes human rights theory for its static conceptualization. His response, with some similarities with Rorty's pragmatic approach in law, is a more open-ended "jurisprudence," a form of case law that recognizes change, experimentation, transformation, historical effects, and metamorphic qualities of being human and of evolving social interactions and institutions. Matza's "drift" theory still remains an exemplary case, that, with modifications by infusion with concepts from the process-informational framework, can contribute substantially to our knowledge of this becoming quality of being human.

5. Cause

a. Classical-Materialist Paradigm (Cause)

Key concepts. Locality; deterministic; clockwork; linear; proportional effects; algorithmic; "God does not play dice"; classical (stochastic) probability and logic.

Commentary. Newton's legacy provides a framework within which cause can be ultimately discovered for any phenomena. The cosmos is made up of fixed particles and fields, operating in a background space of absolute, unfolding time, and the interactions between the two are predictable, or potentially so. We live in a clockwork cosmos one in which determinism is primary. Einstein's quip, "God does not play dice," is the capstone of a materialist ontology of cause. Linearity and proportional effects are givens. Small inputs produce small outputs. Cause unfolds linearly in time and space. A later event cannot cause an earlier event. Cause is local. There must be immediately present determinants for action to occur (locality). Cause is tied to location. Far away phenomena do not have an immediate effect. Ultimately, the cosmos can be reduced to a small set of algorithms, a simplicity that was Einstein's ideal, best example expressed by $E = mc^2$. "Laplace's demon" in contemporary criminology is the empiricist and positivist.[9] The search is for "variance explained." The search is for identifying the determining variables and their respective contributions toward the ideal of 100%. Contemporary studies in criminology (Weisburd and Piquero, 2008) have indicated that on the average we have no better than 40% of variance explained. Thus over 60% is "unexplained variance." More precision, more time, better instruments, more refined conceptualization, more money and surely we will approach the ideal of 100% variance explained. Statistical analysis makes use of classical probability theory whose necessary basis is treating phenomena as if they were more particle than wave.

9. Young (2011: 218) criticizes much empirical approaches in his notion of the "datasaur": "... a creature with a little brain wandering from dataset to dataset, grant to grant; it had a small head yet a huge gut for processing survey material; it was always on the move but it was strangely without direction or purpose."

Thus in determining possible trajectories to crime, probabilities must be assigned, and once an initial starting point is objectified (i.e., seen as a particle), then a deterministic picture can emerge of movement through Euclidean space. For example, Busemeyer and Bruza (2012: 4) point out that in traditional surveys and questionnaires, the first question asked structures a context that will affect subsequent answers.

b. Process-Informational Paradigm (Cause)

Key concepts. Nonlocality; "EPR" ("spooky action at a distance"); non-linear; synchronicity; chance; quantum probabilities and logic; quantum tunneling; "Don't tell God what to do"; holomovement; pilot wave; non-computable; retrocausal; catalysis; QZE; quasi-causal.

Commentary. Cause is not linear, deterministic, and local. It is nonlinear with often disproportionate effects. Dynamic linear systems often have singularities at which nonlinear events are ubiquitous. Small inputs can, with iteration, produce disproportional effects. Chance factors are ubiquitous. Instead of privileging orderly point and periodic attractors, torus and strange attractors[10] populate dynamic systems whereby both order and disorder exist side by side. At best, we have probabilities. Quantum mechanics recognizes this in their notion of a "probability wave." As we earlier discussed, we have "smears" or "clouds" of probabilities. Schrödinger's quantum equations of motion show how these probability waves exist with multiple possible expression. Blood (2009) has well stated the point: "the Schrödinger equations determine the shape of the cloud, how it moves through space and how it responds to other clouds." They remain virtual until their collapse. It is the collapse of the wave function that instantiates one possibility out of the many that may exist in potentiality. Bohr responded to Einstein's famous quip that "God doesn't play with dice," with "Don't tell God what to do." Events only have a range of predictability. We can never hope for 100% variance explained. Ultimately, as Penrose (1994) has suggested, cause is a non-computable event. No "Turing machine" can be created providing complete explanations. The process-informational paradigm makes use of quantum probability and logic,

---

10. A point attractor is the point to which some dynamic system tends. It is its ideal resting place. A periodic attractor, or cyclic attractor, is a recurring pattern; a torus attractor, represented as a doughnut-looking object suggests both determinacy and indeterminacy—determinacy by the limited outlines of the torus, indeterminacy within any slice of the doughnut within which continuous variation prevails; and strange attractors, in their early formulations look like two butterfly wings, each representing a possibility, with only a small perturbation shifting an outcome from one wing to the other.

Figure 1. Classic and Quantum Trajectories

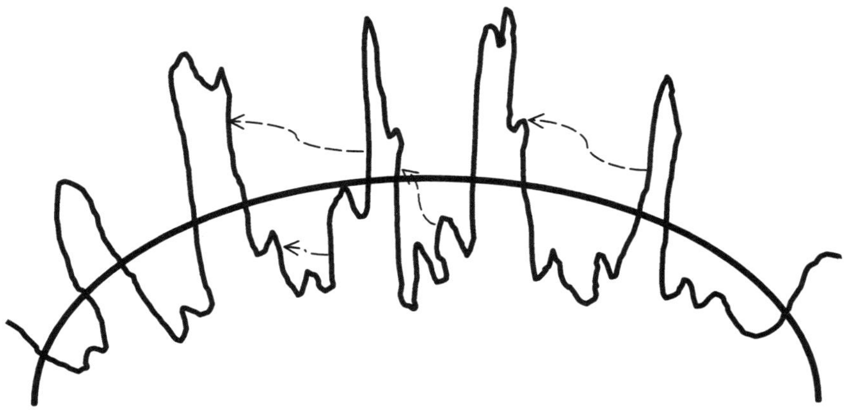

Note: Solid arc represents classics trajectories; crossing jagged represent quantum trajectories; reverse dotted arrows represent quantum backward time referral.
Source: Adopted and modified from Bunge, 1979: 350.

opposed to traditional classic probability theory. In this view, phenomena are in a state of superposition at any moment, that is, are in states of potentiality with various possibilities of instantiation. They are more wave than particle. As Busemeyer and Bruza (2012: 3) note, "there is no single trajectory or sample path across time before making a decision, but instead there is a smearing of potentials across states that flows across time." Arriving at a decision is equivalent to the collapse of the wave function, transforming it into more particle-like (ibid.). And Bunge (1979: 329), in his much-cited book, *Causality and Modern Science,* says "it is not possible to predict with certainty what values a given variable will assume upon measurement." Thus any linear trajectories through time cannot capture the underlying dynamic wave state of all matter and energy (see Figure 1). Measurement (e.g., observations such as surveys and interviews) "create rather than record" some aspect of a dynamic system (Busemeyer and Bruza, 2012: 3). Thus, from Figure 1, classical trajectories are perceived as linear developments in time, whereas quantum trajectories indicate, at any moment, a cloud of possibilities in movement. The collapse of the wave function that describes this "cloud" instantiates one "reality" rather than another possibility. Further, variations of the double-slit experiments (i.e., "delayed choice," "quantum erasure") suggest that once a decision has been made previous information leading to an alternative decision are erased as information.

Einstein, in his much-cited "EPR paper" (1935, co-authored with two colleagues), criticized quantum mechanics for being "incomplete," but did acknowledge the possibility of "spooky action at a distance." Far away phenomenon may have immediate consequences, even defying the upper speed limit of light itself. This is the notion of nonlocality. The cosmos is interconnected and anything happening in one sector can have effects on faraway places. Carl Jung (2010) in psychoanalysis recognized this as "synchronicity." Quantum theory has suggested quantum tunneling as prevalent in the cosmos by which interconnections are ubiquitous and efficacious. Quantum theory has also suggested the notion of retrocausality.[11] Quantum information can be referred backward in time to effect the present. Rather strange to conceptualize given our rigid conceptualization of linear time and causation; nevertheless, the mathematics of quantum theory allow for this phenomenon.

De Broglie (1924) has also suggested that the wave function is not just mathematical, but is physical, a notion that Bohm (1983: 76) built upon in his idea of a "pilot wave," arguing that it represents "hidden variables," holographically encoded information, that "determine the details of its fluctuations" (ibid., 78). This certainly, as Bunge (1979: 348–50) points out, reduces inherent uncertainty, and brings it more in line with more traditional classical principles of causation. The notion of "hidden variables" remains much debated within quantum theorizing.

An alternative to linear causality has been provided by Delanda (2002, 2006) in his notion of "catalysis." If we assume that social entities are assemblages in form (to be further developed in Chapter 3), then one also needs to consider subjective internal organizations. Because the internal organization of an entity is always in process, then each has a differential capacity to effect and be affected as Spinoza and Deleuze have argued (see Hardt, 1993). The same external "cause" will not have the same effect. The threshold point for each entity (as an assemblage) can vary and thus different "triggers" can also have the same effect (see also Delanda, 2002: 140–141, 155). Triggers, too, can be linguistic or nonlinguistic, the latter can be seen in the work of Max Weber on "expressions of solidarity, legitimacy or prestige" (Delanda, 2006: 22). Thus, Delanda argues, reason and motive are important factors in the catalytic causation. Left unanalyzed, however, is the nature of the subject, the person. Even though, as Delanda argues, viewing the person as an assemblage is quite useful, it nevertheless does not provide us with any specifications of the person's underlying inter- and intrasubjective mechanisms. However, Delanda does demonstrate how a nonlinear causal schema has greater purchase in the cosmos. All too often, we linearize

11. For empirical investigation and discussion, see the anthology by Sheehan (2006).

otherwise nonlinear events (Delanda, 2002: 153). Our appetite for the search for 100% variance explained leads to the "accumulation of linear models at the expense of nonlinear ones" (ibid.). Consider Ian Stewart (cited in Delanda, ibid.): "[I]f you decide that only linear equations are worth thinking about, self-censorship sets in. Your textbooks fill with triumphs of linear analysis, its failures buried so deep that the graves go unmarked and the existence of the graves goes unremarked." The tendency, then, becomes one of subsuming problems to given solutions (ibid.).

Apart from the alternative of catalytic causality, Deleuze and Guattari (1987) have also offered the idea of an "abstract machine," a logic that emerges in historical conditions and their constellations of political economic forces. This logic, once created, becomes a "quasi causal operator." For example, consider Foucault's (1977) discovery of the mechanism of panopticism and its asymmetrical form of surveillance and his disciplinary mechanisms, forms of synchronizing bodily forces. These are mechanism of "capture" by which diverse forces of the body are synchronized toward some ideal. Think too of Pashukanis' (2001) analysis of the homology (similarity of development) between commodity fetishism and legal fetishism. Starting from Karl Marx's first 100 pages of *Capital*, much of which builds on Hegel, he shows that the juridic subject, principles of law, and formal equality are derivatives of capital logic. These are, then, abstract machines which are quasi causal operators with effects in the socius. But we cannot view catalytic causation as a linear process. It is the principle of genealogical investigation, originated in the works of Nietzsche and Foucault and further developed in the work of Deleuze and Guattari (1987) in their notion of nomadology that provides a conceptual tool for studying nonlinear forms of causation (this will be further developed in Chapter 6).

c. Applications to Criminology (Critical Commentary, Cause)

Elsewhere, Einstadter and Henry (2010: 12–18)[12] have identified four forms of causality expressed in criminology: linear, multiple, interactive, and dialectical. *Linear causality* is where variables are arranged sequentially, each contributing or working their way through the one following until the dependent variable is explained. *Multiple causality* has two versions. One is where a number of independent variables on their own can produce a dependent variable (crime or delinquency). A second, where a number of interdepend-

12. DiCristina (1995) identifies nine forms; see also, Henry and Milovanovic (1996: 125–127).

ent factors exist where any one by itself is insufficient in producing the final event (crime or delinquency). *Interactive causality* is where any outcome can have retrocausal effects; that is, the outcome can feedback on the events that had proceeded, which, in turn, have effects on the outcome.[13] This has been conceptualized as "reciprocal causes," "bidirectional causes," and "causal loops" (Empey and Stafford, 1991: 288–89; DiCristina, 1995: 6; Thornberry, 1987; Thornberry and Krohn, 2005; Sampson and Laub, 1993: 245, 249). In this scenario cause and effect are influential on each other: there is a cyclical movement from forward to backward to forward. A related version to interactive causality is the social network perspective (Fischer, 1982; Krohn, 1986) that looks at the structural relations among people that produces constraining effects irrespective of the characteristics of individuals. *Dialectical causation* or *co-determination* is a form of multiple interactive causation (Henry and Milovanovic, 1996: 126; Einstadter and Henry, 1955: 15–18). There are three components (Henry and Milovanovic, 1996: 126):

> First, the cause and outcome are not conceived as discrete entities but are interrelated and overlapping, such that some part of cause is constituted by some part of the event produced in part by it and vice versa … Second, … neither 'cause' nor 'event' has causal priority since each has simultaneous transformative powers over the other … Third, changes in either the 'cause' or the 'event' are not temporarily separated because their interrelatedness means that as one changes, the other must change, and with it the first must change also. Reciprocity of influence then is instantaneous, although cumulative mutual transformation may extend over time.

The dialectical or co-determination approach, as well as interactive causality with more, are accommodative to quantum theory. First, neither contributing entity nor final event are discrete, static events but are instantiations of the quantum wave function. In other words, each variable changes in values at any moment in time. Second, neither contributing entity nor final event has primary causal efficacy since each can have a simultaneous effect on the other. This is consistent with the idea that an assemblage can have differential powers to effect and be affected (catalytic causality) and that retrocausality renders linear explanations insufficient. Third, the contributing entity and final event are not temporarily separated; each affects an instantaneous change in

13. This would also map to the notion of backward referral of quantum information with, in turn, effects on the present.

the other, a form of nonlocality. In each case, it is information and information exchange that is critical in how relations unfold in process. This is a process-informational perspective.

Contemporary classical linear and multiple causation driven criminology often make use of various forms of path analysis (*multiple causation*). This is an impressively looking spatialization of the causal links with quantified contributions of determinants (independent variables) in various configurations flowing in a linear clockwork time frame (via intervening variables) producing a particular dependent variable. (This is suggested in Figure 1 in comparing classical trajectories with quantum trajectories.) Each contributing variable is rendered static, a snapshot in its otherwise ever changing state. The end result is a flow chart of how crime or delinquency takes place. It also assumes point and/or periodic attractors. Dynamic systems theory would suggest that we should be more occupied with nonlinear effects, looping, feedback, and various attractors that may characterize it (e.g., torus and strange attractors). Catalytic causality would suggest that each of the states of variables caught in snapshots should be seen in their differential abilities to effect and be affected.

6. Information

a. Classical-Materialist Paradigm (Information)

Key concepts. Digital; bits (Shannon); binary; classical holography; Boolean; two-logic; brain as receptacle; axiom-premise-syllogism-truth; reduce problems to solutions.

Commentary. Information and its processing, coding, storage, transmission and decoding are concerns of both paradigms.[14] Shannon's (1948; Shannon and Weaver, 1998; Pierce, 1961) classic research on bits of information has become the center piece of our digital age. Computers, digital recorders, television, and cell gadgets—all rely on Shannon's basic informational principles. The classical-materialist perspective more often relies on Boolean logic, yes-no switches (a two-logic) that can be quantified and plugged into syllogisms in arriving at various "truths." This is exemplary in law. "Normal science" tends to "subordinate problems to solutions" (Delanda, 2002: 144). That is, given a problem the search is more often for invoking a pre-established stock of solutions. Lefebvre (2008) has argued this idea also pertains to judges'

14. There has been much recent discussion on the Web as to whether the universe is digital or analog.

decision-making whereby a constructed "case" is subsumed ("subsumption") under a given legal concept, thereby allowing a syllogism or linear logic, to play itself out toward a conclusion in law. Consider Aristotelian logic that is the basis of syllogistic reasoning/deductive logic: all desire is good (major premise) → ambition is desire (minor premise) → therefore, ambition is good (conclusion). In this model, truth is to be discovered; it already exists objectively within the boundaries of contemporary, historically situated society. The brain, in this paradigm, is a receptacle of information, the latter stored as engrams in constellations of neurons. Sigmund Freud, in *Interpretation of Dreams* ([1900] 2000), developed one of the early comprehensive integrated approaches linking how primordial sense data are stored in constellations of neurons called "thing presentations." Electrochemical pathways determine how information flows within the brain and body. In this paradigm, transmission of information and interpersonal communication takes place directly via a common, neutral language system.

b. Process-Informational Paradigm (Information)

Key concepts. Analog, logons (Gabor), quanta of information; qubits; 2-D nested boundary surfaces; quantum holographic; 4-logic; topos; vortices; holographic principle; implicate order; pilot wave; noosphere; biocomputer; Wheeler, primacy of information; "in-formation field"; problematic approach/appropriate questions versus normal science.

Commentary. Information is patterned, organized energy. It is geometrically encoded as frequency interference patterns (holograms) stored in the holofield, or ZPF. It is a result of ongoing dynamic processes. Each process is information construction and exchange. A number of prominent theorists have argued that information and information exchange is not secondary, but primary (Wheeler, 1990; Smolin, 2001; Bekenstein, 2003; Verlinde, 2010). Consider Wheeler's (1990) conclusion that we should see the "physical world as mode of information, with energy and matter as incidental." More recently, Verlinde (2010) has demonstrated that the laws of Newton (as well as Einstein's principles) can be derived from information theory, thus substantiating the view of the primary of information and information exchange. Critical to an understanding of information is how it is encoded in wave form. A good visual example is provided by Laszlo (2007: 70–71) of ships plying the seas. (Or, consider sitting by the side of a pond throwing stones into it, experiencing how waves interact.) They leave wakes which interfere with each other producing complex patterns. This modulation of the sea's surface produces information about the ships. "It is possible to deduce the location, speed, and even the tonnage of the vessels by analyzing the interference patterns of the waves they have created" (ibid., 71). Each new ship produces a new modulations, new information. Electromagnetic waves are carriers of information. A stream of photons impinging on our senses carries information. A wave is composed of different amplitudes (height) and wavelengths; a single oscillation is a cycle. The frequency concerns the number of cycles per second. Two waves are in phase when their peaks and troughs are congruent. (We shall extend on this notion of being in phase or resonance with the idea of phase conjugation in the next chapter.) The sine and cosine waves are the most familiar ones. Phase waves are called interference patterns which are also stronger than each of the parts. The interference pattern carries information of the component waves in one momentary composite wave. Gibson (1979), for example, has shown that ambient light carries information. A number of theorists have shown that all entities emit waves, or encoded information, as well as absorb them.

Information storage has been seen differently than from the classical view. Information storage has been explained quantum holographically in several ways: within the brain (Pribram, 1971, 1991); within the zero point field (ZPF)[15] (Laszlo, 1995, 2007; Puthoff, 1998a, 1998b); and encoded on boundary surfaces of a volume of space (Thorn *et al.*, 1986; 't Hooft, 1993; Susskind, 1995, 2009; Bekenstein, 2002; Maldacena, 1998, 2005; Bousso, 1999, 2002). Information is encoded holographically (Gabor, 1946) and is nonlocally accessible. We often recognize a hologram with a credit card logo or in Princess Leia's image in *Star Wars*. It is constructed by the interference of two waves. One, emitted by an object, is combined with a non-object bearing wave which produces an interference pattern that can be encoded on a recording surface (much like in the double-slit experiments). On this surface it appears as merely randomly scattered lines and scribbles. Direct the same non-object bearing wave on any part of this surface and the initially encoded three-dimensional image will reappear. The holographic image's unique characteristic is that it is spread out all over the recording surface (nonlocality) and activation of any part will re-produce the whole image. A hologram can store a huge amount of information. One sugar-cube-sized hologram could store all the information of the U.S. Library of Congress. The concept of noosphere (Deleuze, 1989) suggests that information is bounded and ecologically specific.

Information is primarily stored analogically, not digitally, in the form of a quantum bit, or "logon" (Gabor, 1946), also referred to as a quantum of information. A logon is the smallest unit of discernible information and is en-

15. See also related discussion of zero point energy by Einstein and Stern, 1913.

coded holographically (ibid.). Hence, this is a quantum hologram as opposed to a classic hologram. What makes it quantum is that there is an inherent indeterminacy about the logon. Gabor (1946), applying Heisenberg's uncertainty principle, shows how the logon is constructed along the same two primary dimensions of momentum and location. He also indicates that logons overlap other logons and even posits an "overlap with the future." That is, inherent in any logon is an anticipatory dimension. Classical holograms, on the other hand, such as suggested by the earlier work by Laszlo (1995) and Bohm (1983), especially with the latter's notion of an "implicate order" and "pilot waves," are self-enclosed and thus in accessing them implicated is a deterministic outcome, a form of holographic determinism. Current computer technology makes us of digital, on-of switches. The newly emerging quantum computer will make use of qubits, or quantum coherent states of both "yes" and "no" during which computations are conducted. A collapse produces a distinct answer to a question posed.

Classical information theory assumes a 2-logic: answers to questions posed are ideally conceptualized in the format of "yes" or "no." Quantum theory, on the other hand, embraces a 4-logic (Rauscher and Targ, 2001). Contrary to classical thought, Eastern Philosophy has been shown to embrace a "four-logic system" (Garfield, 1995; Hayes, 1994). Here, narratives about the cosmos can be true, not true; but also, "... both true and not true," and "neither true nor not true." Applying this to quantum mechanics, Rauscher and Targ (2001: 344) explain that "light is (1) a wave, (2) not a wave, (3) both a wave and not a wave, or more correctly, (4) neither a wave nor not a wave." Thus a two-logic or a four-logic may reflect, alternatively, key assumptions of the two respective paradigms. There is compatibility here with the Greimasian square[16] used as a mode of discovery of "deep structures" in the human lexicon. Here there are four corners. The upper left and right are contradictories, such as "yes," "no." Diagonally to "yes" we have "not yes"; diagonally to "no," we have "not no." A "not yes, not no" is compatible to quantum states (e.g., Schrödinger's cat in the box being both alive and dead until observation; an entity is both wave and particle).

Information is in-process, not static. We saw this in the quantum erasure experiments. Observations and measurements create not just discover and record information which has in turn corporeal effects on the organism. Clas-

16. See its diagramed form: http://en.wikipedia.org/wiki/Semiotic_square. See also applications of the semiotic square to law (Milovanovic, 2003: 198–203); to the use of the insanity defense and the right to refuse treatment (Arrigo, 1999); to legal discourse (Greimas and Landowski, 1990; Jackson, 1985, 1995; Arrigo, 1998); to the dialectics of true and false and guilty and not guilty in criminal trials (Jackson, 1998); and, generally, to jury verdicts (Jackson, 1998).

sical physics assumes that information can be discovered as knowledge; it is said to already exists in the cosmos. The process-informational approach sees information and knowledge as ongoingly created.

c. Applications to Criminology (Critical Commentary, Information)

Information in traditional criminology is a static construction. For empiricists, ever more refined constructions in operationalization of variables and in developing better models for variance explained assume that there is an objective, static truth/knowledge in existence. Constitutive theory in criminology (Henry and Milovanovic, 1996) although an advance on traditional criminological theorizing, does suggest the interconnectedness of information construction between the more macro and micro levels. But it is in further need of indicating how creative change takes place (see Delanda's critique, 2006). Quantum holographic theory is suggestive of a non-deterministic view of constitutive theory. For traditional participant observer methodology, it is assumed that a degree of objectivity will provide better access to information and truth. However, we have already witnessed that the very observation/measurement by the researcher creates reality by collapsing the wave function, often effectively transforming the subject s/he studies. Yet another area where there are polarizations is in the establishment of "truth." Contrary to classical thought that assumes that observers can ultimately decide on the truth or falsity of a statement, Smolin (2001: 28–32) has argued that truth and falsity of a statement are observer dependent. Since different observers have access to different information, he argues, "we almost always reason with incomplete information." He does suggest an alternative, an "intuitive logic" in the form of "topos" theory. Similarly, Deleuze (1989), following Nietzsche, argues that there are two competing regimes or background spaces (noospheres/noosigns) from which information in the form of "truths" are constructed. One form is where truth is assumed "to be achieved, formed, or reproduced" (ibid., 146); it is to be discovered as a constant in the real world. It is a system of judgment from which the "man of truth" is the exemplar. "He holds up a superior value, the good, in the name of which he will be able to judge, he is craving to judge, he sees in life an evil, a fault which is to be atoned for" (ibid., 137). Two forms are the man of revenge who tries to actively correct her/his constructed wrong (think in terms of the vigilante) and the sick man, a person of ressentiment who spirals ever further into his/her nihilism (ibid., 140–141). The second form of information and truth construction is derived from the "power of the false," an ongoing questioning of fundamental presuppositions, a suspension of the taken-for-granted world, a falsifying outlook acknowledging all information is in-process, best exemplars being an artist, a creator of truth, and the "man of generosity" (ibid., 147). Here, information and truth are created; they are constantly in process, mutating, transforming, defying fixity, opening up ever new possibilities. In traditional classical criminology the emphasis and background space (contrast space) is on the former type of judgment. Indeed, we see the "man of truth" idealized in videos, fictions, law, etc. We shall return to this in Chapter 6 on developing suggestions for an alternative, transformative justice.

7. Language

a. Classical-Materialist Paradigm (Language)

Key concepts. Static; categorical; noun privileged; subject-verb-object; neutral instrument/medium; words as isolated, distinct separate entities; locality; "spreading activation model"; linearity.

Commentary. Language for the classical-materialist paradigm is seen as a neutral medium by which communication is possible. Dominant Western languages privilege the noun form over the verb form. The basic structure is: subject-verb-object. Implicit is a subject who is determining, a verb which reflects the action of the subject, and an object which is what the action is about. We, as authors, choose our words, or signifiers, from the language within which we are socialized. These are noosigns circulating within a noosphere. Benjamin Whorf, in *Language, Thought and Reality* (1964), explains how a language carves up "reality" in a very distinct way. Language is static in as much as dictionary meanings dictate the appropriate usage of signifiers. A linguistic code structures the correct construction of narratives in producing meaning. Daniel Chandler's *Semiotics: The Basics* (2011, and on-line) reviews how two axes, paradigm or word choice, and syntagm, or the semiotic code for correct grammatical constructions, together organize narrative constructions. The dominant theorists that inform much of western thought on semiotics include Ferdinand de Saussure, Charles Sanders Pierce, and Algirdas Greimas. A psychoanalytic semiotics has been developed by Jacques Lacan (1977, 1991) who favors de Saussure's versions of linguistics. We will also show in the following section, that a quantized Lacan is a productive line of new inquiry.

b. Process-Informational Paradigm (Language)

Key concepts. Dynamic; language wave function; context/cues collapse wave function; polysemy; linguistic relativity; verb privileged; quantum holographic lexicon; quantum mental lexicon; word entanglement; non-separable; relational; rheomode; genetic texts; "spooky activation at a distance model"; nonlocality; semantic spaces; wordspace as vectors in Hilbert space; holographic reduced representations (HRR) and BEAGLE models; associative holographic memories; quantum cognition.

Commentary. Language is conceptualized in terms of a wave function. Two directions exist: a quantum mental lexicon (Nelson and McEvoy, 2007; Bruza *et al.*, 2009a, 2009b; Bruza and Kitto, 2008; Bruza *et al.*, 2008; Kitto and Bruza, 2011; Newman, 2008); and a holographic lexicon structure (Jones and Mewhort, 2007; Cox *et al.*, 2011). A mental lexicon concerns words and their associations. Quantized language is also a component of the emerging discipline of quantum cognition (see the journal *NeuroQuantology* established in 2001; Busemeyer and Bruza, 2012); we will return to this in Chapter 6. Quantum approaches depart from classical thought in suggesting that words are relational, inseparable, part of a network, nonlocal in activation, and quantumly entangled. An initiated word simultaneously mobilizes other associated words ("spooky activation at a distance model") unlike classical thought which argues that a distinct word linearly activates other words ("spreading activation model").

Signifiers exist in entangled states and only in context do they undergo a collapse, a collapse of the wave function. Signifiers are polysemous (Newman, 2008); they have many potential meanings (Bruza *et al.*, 2008). A word, or signifier, using quantum notation is represented as a wave function, $|\Psi>$, or |word>; it's "basis" would then be: $|\Psi_i>$, where "i" represents potential meanings, or signifieds. Signifieds remain quantumly entangled. A given context is equated with choosing a particular measurement/observation (Bruza *et al.*, 2008, 2009: 7, 9, 13–14), which collapses the wave function; the many possible states reduce to one actualized. As Kitto and Bruza (2011) state, "a word changes from an extended to a local meaning state when encoded in a context." This has compatibilities with a quantized Lacan (1977) in so much as the latter recognized the natural state of a signified sliding under a signifier until a "punctuation point," rendered as s(O), is attained, at which point, a signifier is attached to a particular signified. A quantized Lacan also has compatibilities with the idea that speech production entails two temporalities: a forward, anticipatory dimension which would find support in Gabor's notion of the logon, and a "retrograde" or retrocausal constitutive role which grasps the unfolding chain of signifiers all at once producing a distinct meaning of the narrative construction. Now momentarily stabilized, strung together signifiers take on specific individual (relational) value, and together produce a meaningful statement.

Whorf's (1964) classic study identified the Hopi Indian of Southwestern America as best representing a quantum language, in so much as the verb form is dominant; the "I" as the determining subject almost non-existent; and a view of becoming and action interconnected at many levels of the cosmos. He also suggests that any given language predisposes its users in carving up "reality" in one way rather than another form of expression. Bohm (1983), in a separate development, noted how the very language scientists use predisposes their formulation in the static direction; he suggests, as an alternative, a verb-oriented language, a "rheomode," which resists static constructions.

The second form of a quantum lexicon is the holographic quantum lexicon (Plate, 1995; Jones and Mewhort, 2007; Cox *et al.*, 2011; Hannagan *et al.*, 2011; De Vine and Bruza, 2010; Kelly *et al.*, 2013). Several models have been developed: the "holographic reduced representation model" (HRR; Plate, 1995; Kelly *et al.*, 2013), building on Gabor (1969), shows how human lexicon can be modeled by a holographic associative word/memory structure; and a derivative, the BEAGLE model ("bound encoding of the aggregate language environment," see Jones and Mewhart, 2007; Cox *et al.*, 2011), which has the ability to explain how both words *and* contextual information are encoded by the use of vector analysis in Hilbert Space (e.g., a multiple dimensional vector space). Words, then, would be represented by frequency patterns and waveforms (De Vine and Bruza, 2010; Jones and Mewhart, 2007). Holographic quantum lexicon analysis by these theorists which assumes nonlocality could, by this very logic, extend beyond the enclosure of the brain as repository of memories and words, as assumed in Bruza, Kitto, McKevoy and Nelson.

Explanations have appeared in the literature that demonstrate how a noosphere with distinct noosigns are historically stabilized as a holographic lexicon (Jones and Mewhart, 2007). Relatively stabilized signifiers can be conceived as "invariant structures" (Gibson, 1979) and provide "cues" or "affordances" (distinct possibilities) to those within the noosphere. We will have occasion to develop this further in the next chapter.

c. Applications to Criminology (Critical Commentary, Language)

Classical-materialist inspired criminology assumes a static language. Identities in the criminal justice system, legal categories, rights, identity politics, and justice principles all are static blocs of spacetime devoid of process and becoming. Deleuze (2004) opposes rights because of their static construction. He suggests a more dynamic, ever-changing "jurisprudence," closer to case law. Woolford (2009) has argued that identities in criminal justice and restorative justice (victim, offender, community, facilitator) are discursively constructed and reified in their continuous use. They need to be deconstructed, he argues, if we are to move forward to more meaningful transformations of the justice system (we shall have more to say in Chapter 6). Identity politics, a search for "roots" and their revival, without more, has been criticized for its static dynamic (Hardt and Negri, 2005; Deleuze, 1989; Rodowick, 1997). Offered, instead is the notion of a "people yet to come," identities that draw both from the past, the present and toward a novel future which incorporates what the body is capable of doing. This would entail the development of new signifiers that are more likened to a wave function indicating many possibilities of expression. In one of the few approaches recognizing a process-informational ontology in criminology, Matza (1961) has shown how delinquency is a process involving a "drift," instantiations of clouds of possibilities in particular form in context, in an otherwise state of perpetual movement. It is often societal reaction that is the catalyst in the collapse of the wave function producing particular static identities which then recursively generate a continuation of delinquency. The process-informational ontology provides alternative conceptualizations with which to grasp not only the physicality of these processes and opportunities for specification, but also for transformative practices that, following a Spinozean ethics, increase the person's capacities for effecting and being affected toward a realization of the fuller potentials of being human. In Chapters 5 and 6 we will also explain how ecological niches, enclosed subcultural domains, and other bounded spaces produce affordances, lines of flight, transmitted by ambient light as holographic information, that interact with the subject's unique quantum signature wave in reproducing scripts that are conducive to delinquency. Particular recurring collapses of the wave function perpetuate a narrow band of core master signifiers ("order words," Deleuze and Guattari, 1987) and signifieds for instantiation in everyday discourse from which reality construction takes place, ongoingly reifying the dominant order. We will also offer suggestions for a transformation justice which follow the ontology of the process-informational paradigm.

8. Processor of Information

a. Classical-Materialist Paradigm (Processor of Information)

Key concepts. Neural networks; bio-chemical; brain, neurons, synapsis, dendrites; visual correspondence; representational; neuronal subject; neurophilosophy; neural darwinism; snapshot science.

Commentary. Informational processing in standard texts depicts a person as a composite of neural networks extending from the brain to every extremity.

The five sense organs receive stimuli and process them internally, all coordinated by brain activity. Internal coordination is by way of messages delivered by bio-chemical transmission processes, all coordinated to produce almost immediate responses to incoming stimuli. Consciousness is fundamentally reducible to the activities of coordinated neurons and their firing activities (Koch, 2004; Koch and Hepp, 2006, 2007; Gazzaniga, 1985, 2012; Crick, 1994; Crick and Koch, 1995, 1998, 2002; Ryle, 2002). Sigmund Freud's classic *Interpretation of Dreams* (1900) explained how incoming stimuli are initially registered in neuron clusters within the brain in mnemic systems. Each cluster remains embedded with specific quanta of energy that when activated by the work of various unconscious tropes (condensation, displacement, secondary revision, and considerations of representability) produces the thoughts and images that constitute our reflective realities. How precisely an initial inscription (registration) of a sensory image takes place, how it is then reinscribed in the unconscious, how, further, it becomes a "thing-presentation" (signified) to which a "word-presentation" (signifier) is connected can be better understood by the "mapping to and from" dynamics of phase conjugation to which we will return in Chapter 4.

Several classical-materialist approaches have pushed the logic of neural connectivity to the extreme (for a particularly concise summary, see Germine, 2008). They include: a "neural Darwinism" whereby neurons are organized according to the principle of "survival of fittest" (Edelman, 1987); "neurophilosophy" (Churchland, 1986) which advocates for the neural correlates of consciousness; and the assertion that free will is illusory (Dennett, 1991; Wegner, 2002), perhaps best expressed by Ryle's (2002) idea of the "ghost in the machine," or even Huxley's (1986) emphatic, "we are helpless spectators."[17] Perhaps Koch and Hepp (2006, 2007) are exemplary in their continued criticism of quantum approaches. They argue, instead, for classical principles as the basis of higher brain functioning. Crick (1994) and Edelman (1987) independently have received Nobel prizes for their work in this area.

Sensory data, in this view, begins from outside of the body, registers on the sense apparatus (visual, sounds, taste, smell, feel), undergoes Fourier transformations, which then mobilizes the body's bio-chemical and neural structures to respond in a coordinated fashion. For example, light rays emanating from an object enter the eye, are converted into frequency waves via the mathematical principle of Fourier transforms and are then constructed in the back of the brain as three-dimensional objects. Fourier transforms are mathemat-

17. See also Chalmers (1996).

ical ways of converting the "out there" into wave frequency patterns, said to take place by the eye. The image is internally constructed. This is a representational dynamic that assumes a one-to-one correspondence with the object and image constructed in the back of the brain. Images constructed are then provided appropriate linguistic labels which tend to become stabilized in their use. Criminology more likely takes these labels to be true representatives of objects and organizes them into their categories for explaining crime. This is a tendency toward a snapshot science by which static blocs of spacetime are produced and reconstituted in their everyday use.

b. Process-Informational Paradigm (Processor of Information)

Key concepts. Quantum holography; holographic transducers (emitter/absorber of quantum holographic information—DNA, mitochondria, microtubules, liquid crystals, heart); frequency analyzer; imagemaker; phase conjugate adaptive resonance (Pcar); psychophysiological coherence; Fourier transforms; biophotons; holonomic; bioholography; quantum cognition.

Commentary. We begin with the provocative statement by Bergson, *Matter and Memory* ([1896]2002). He likens the brain to a form of central telephone switchboard (ibid., 30) that can delay or allow communication. The brain, too, is an image amongst a universe of images with the unique capacity to delay immediate response (ibid., 19). It is a "center of indetermination," a "living image"; it acts like other images "receiving and giving back movement," but with the added capacity that it can delay immediate response during which perceptions, the basis of subsequent action, can be constructed (more on this in the next chapter). More recently, the processor of information has been seen in terms of quantum holography. Every entity is vibratory by nature; it has a unique signature frequency. The entity both absorbs and emits electromagnetic radiation. Sense organs are conceived as transducers or frequency analyzers that emit and absorb this radiation. The brain, and so too the heart, are emitters of relatively strong frequency radiation (McCraty *et al.*, 2004; McCraty *et al.*, 2009; Bradley, 2006; Bradley, 2009). Biophotons (Gariaev *et al.*, 2001, 2011; Popp, 2000, 2012; Rahnama *et al.*, 2011; Bajpai, 1998; Chang *et al.*, 1998) are produced by the body and are encoded with holographic information. Pribram's (1991) revolutionary finding was that the brain operates holographically (i.e., "holonomic" brain); it is likened to the recording plate on which interference patterns are recorded and from which, three dimensional images can be reconstructed. Uniquely, in this view, information is spread all over the brain (nonlocality), not stored in particular clusters of neurons separate from others. In this view, the brain is a transducer which accesses encoded information from this ambient field and via forward and reverse Fourier transforms constructs three-dimensional images projected externally. The brain has its own unique quantum signature wave which provides differential access to this information. (We shall return to this in more detail in Chapter 4 and subsequent chapters.)

The key mechanism by which images are first transformed from the world into wave forms (Fourier transforms) and subsequently, via a reverse Fourier transform reconstructed in three-dimensional image form is phase conjugation (Marcer and Schempp, 1997a, 1997b, 1999; Mitchell and Staretz, 2011). A process called Pcar, phase conjugate adaptive response, is offered that parallels in many remarkable ways anticipatory statements by Bergson in 1896. Recall, for Bergson images are constructed out there where they are located, not in the brain.[18] An emitting entity's waves are received by the person's sense organs which return a signature wave along the same pathway back to the entity highlighting (collapsing) certain features. The original emitting entity then reflects this image back to the receiver who now acts on this constituted image. It is phase conjugation, or resonance, or better, an interference pattern that is created between the receiving and sending entity that constitutes "objects" in the form we perceive as reality. In this view, consistent with Bergson, consciousness is *in* objects as distinguished, for example from phenomenologists, who claim consciousness is *of* something. Some, therefore, have characterized the person as an "imagemaker" (Marcer and Schempp, 1997a).

Whereas Pribram located a holographic image-making process within the confines of the brain, others have located the process externally as suggested by Bergson. Each person's unique quantum signature wave accesses either the zero point field (Laszlo's in-formation field, 2007) in one view, or in a more radical version, the enclosed surface on which all information is stored of happenings that enclose this volume ("holographic principle"). We interact within bounded spaces and the enclosed "surface" on which holographic information is imprinted. We will have more on this in the next chapter.

DNA, in this model, acts as a holographic projector/transducer (Gariaev, *et al.*, 2001, 2011). "DNA is a self-calibrating antenna working by phase conjugate adaptive resonance capable of both receiving and transmitting quantum

18. An earlier model has also been developed, a "transactional interpretation" approach, by John Cramer (1986, 2001, 2006a, 2006b; see also Elitzur and Dolev, 2006). It has been acknowledged (Marcer and Schempp, 1997) as "compatible" with their phase conjugation model. We focus on Marcer, Schempp and Mitchell for their inclusion of consciousness, whereas Cramer's four step model disallows the effects of consciousness on the collapse of the wave function.

holographic information stored in the form of diffraction patterns ... quantum holograms" (Gariaev *et al.*, 2011: 17; see also the review article by Miller *et al.*, 2002). Emitting radiation from DNA quantum holographically encodes and decodes information which then affects bodily cells and structures by the process of quantum nonlocality. Thus the difficulties of classical thought as to how DNA can communicate seemingly instantaneously has been answered by a form of teleportation or nonlocal dynamics.

Two alternative models of a quantum consciousness have been developed whereby consciousness collapses the wave function. Hameroff and Penrose (1996a, 1996b, 2003) argue that collapse of the wave function takes place "objectively" when a certain energy state is reached and it is traced to the work of microtubules, the skeletal structure of all neurons, responding to momentary breaks in spacetime geometry. This break produces quantum coherence where alternative possibilities exist side by side until the appropriate threshold energy value is reached at which point a collapse takes place and one possibility is instantiated. The second model is by Henry Stapp (2007) who, following von Newmann's (1932) classic statement, traces the collapse of the wave function to the "abstract ego" posing questions the answers to which nature responds in its own way. This produces two collapses, one with the selected question posed, another with nature deciding on the answers to which the ego now responds. We shall return to a further, more detailed examination of the work of Stapp, Penrose and Hameroff in Chapter 4.

c. Applications to Criminology (Critical Commentary, Processor of Information)

Classical materialist-inspired dominant criminology is vacuous as to the wherewithal of an agent, a subject outside of the confines of neural-chemical explanations. Within these confines free will withers away. Determinism in the form of positivism is ubiquitous. The highly valued empirical approach, even as we have indicated previously that averages no better than 40% variance explained, is seen, nevertheless, as the mode of discovery of objective facts through ever more precise measurement tempered by possible threats to internal and external validity. The subject within this framework is not necessary. It is merely more "data" that can be operationalized, or taken as a given within the neural framework. Recent resurrection of rational choice theory, and its variants which posit a volitional subject with unrestricted choice capabilities, offers no bona fide statement about the inter- and intra-subjective constitution of the subject. A good part of critical criminology fairs no better. Fluctuating between understanding deviation because of structural inequities (assumes a deterministic model), attributing major cause to subcultural factors, and the left realists (assume limited choice) who do not romanticize the offender as some kind of incipient revolutionary—in each case, there is no, or a restrictive notion of the subject at work. It is time to rise to the challenge and consider possible constructions that are interconnected more dynamically with political economy. The process-informational paradigm has offered material for a fresh approach. Neither Bergson's nor Deleuze's much innovative theorizing sufficiently answers this call; for both, in their haste to dismiss any form of a "transcendental subject," they have effectively effaced the subject of her/his uniqueness, even as both place a premium on a becoming. Nor do we want to advocate a transcendental subject, for we want to show that the subject is both constitutive of, and constituted by the quantum informational field within which it exists. In Chapter 4 we shall turn to one possible construction, Schema QD, which is a de-oedipalized and quantized version of Lacan's Schema R.

Conclusion

The classical materialist paradigm remains embedded in ontological assumptions in contemporary criminology. Exclusive reliance on this conceptual space reduces our opportunities to think otherwise, to pose alternative questions, to formulate frameworks which remain in-process, responding to the uniqueness of being human in society. We argued that quantum holography theory has emerged, but yet, with few exceptions, most notably in neuropsychology and the study of human lexicon, has not been seriously entertained in theorizing in criminology. It is time for a change. We need to address the incredible failure of our criminal justice system. Our opening chapter is intended to introduce the reader to new possibilities in theorizing in criminology. It has not been intended as an exhaustive listing of all the developments in quantum and holography theory and their possible integration in criminological thinking, but is more in the spirit of Jock Young's call for the expansion of the "criminological imagination."

Chapter 3

Holographic Principle, Primacy of Information, and Assemblages

Introduction

The previous chapter has familiarized us with the differences between two paradigms, the classical-materialist rooted in Newtonian physics, and the process-informational paradigm rooted primarily in quantum and holography theory. In this chapter we want to go into more detail on four critical concepts in the latter approach: the primacy of information; communication theory rooted on the notion of phase conjugation resonance; the nature of entities and the dynamics of their interactions derived from Deleuze and Guattari's work on assemblages and abstract machines; and the notion of quantum holographic ecological niches. Two exemplary applications from the social sciences will be summarized. Our last section will apply these various concepts in showing how they work in criminology. In the next chapter we will tackle the "hard problem" of the wherewithal of consciousness. We will develop the notion of a quantum holographic based consciousness/agency.

Quantum Holography and Primacy of Information

Quantum holography has offered novel conceptualizations of how information is constructed, stored, and transmitted. Holographic theory has had four significant developments: one beginning with the work of Denis Gabor (1946) on message transmission; a second, on the nature of an "in-formation" field and its access by way of phase conjugation (Laszlo, 1995; Marcer and Schempp, 1998, 1999; Mitchell and Staretz, 2011); a third by Bohm (1983;

Bohm and Peat, 1987) who suggests an implicate and explicate order; and a fourth derived from the study of black holes and their event horizons.

Conventional science offers a correspondence approach in construction of "what's out there." Spacetime patterns (the "out there") are transmitted to the retina which performs a Fourier transform, converting images into frequency patterns and electrochemical impulses which travel along the optic nerve that then are decoded in the back of the brain in the vision center. It is here where we actually see the image.[1] There is correspondence between the external image and the image that one perceives in the back of the brain. This process has been fundamentally challenged by approaches in quantum holography. Let's briefly review four.

1. Gabor's Quantum of Information (Logons)

The first development was by Denis Gabor (1946, 1948, 1969). Gabor (1946) was working on how to improve the transmission of telephone messages across the Atlantic. He based his analysis on Heisenberg's uncertainty principle (one cannot simultaneously precisely measure momentum *and* location). He showed how information is encoded in wave (analog) form in overlapping quantum bits, which he called logons. These are "quanta of information," which encode simultaneously spectra (interference patterns) and spacetime patterns. A logon is the minimal possible amount of a pattern of energy that still allows intelligibility of the information. Said in another way, it is the minimal amount of compactificating or compressing of a pattern of energy to reduce uncertainty in rendering something intelligible. The point at which it becomes unintelligible would be the limit of discernibility. Logons have an indeterminate character. That is, due to the very limitations of precise measurement while in movement, uncertainty remains ubiquitous. In 1948 he conducted research in improving the microscope and was able to show how a holographic principle applies in the optical area.

Karl Pribram (1971, 1991, 2004, 2006, 2009) was influenced by Gabor in developing a holographic model of brain functioning. His approach fundamentally challenges conventional analysis in neuroscience. He met Gabor in Paris during a UNESCO conference (Pribram, 2004: 219) and had an exchange that was pivotal in developing his idea of a "holonomic" brain; that is, brain func-

1. Hypothetically, it is conceivable to hook up the brain to a computer program generating visual images, much like in virtual reality computer simulations, and a person would literally "live" in this world. For a clear popularized example, see Star Trek: The Next Generation (1992), "Ship in a Bottle."

tioning based on quantum holography. Encoding information in the brain was likened to the recording surface in the double-slit experiments (see Chapter 2) which showed how interference patterns are recordings on a receiving photoelectric plate. Information, for Pribram, is stored non-locally in the brain, not in particular clusters of neurons. In his view, spacetime patterns (i.e., experienced everyday reality, or "spacetime observables," Pribram, 2009: 457), which is a frequency domain, are received by the retina, which, by way of Fourier transforms, transforms these frequencies into interference patterns. The latter encode not only the frequency, but also the amplitude and phase. This is the "spectral domain." These are "spectra," one form of which are holograms. Spectra, as interference patterns remain in flux and represent potentials; they are, according to Pribram (2004: 231), "pre-spacetime."

More specifically, Pribram locates these interference patterns in the synapse and dendrite junctures of the brain; they establish criss-crossing interference patterns spread over the whole brain (ibid., 88). It is thus wave interference patterns with which the brain communicates (see McTaggart's clear, accessible expose, 2008: 84–85). He (2004) also noted all our senses are organized in this manner. This information (hologram or interference pattern) is transmitted to all parts of the body in a nonlocal way; meaning, it is instantaneous. It is for this reason that the human can engage in complex behavior in seemingly a coherent, smooth, fast, and orderly manner. In other words, primordially given is a world in flux reflecting energy in movement indicated by frequency patterns. It is from this flux that spacetime coordinates are constructed (2004: 9). This transformation, he (ibid.) informs us, constitutes "material and temporal locations of information and meaning." While we experience the everyday world in spacetime, our brain and sensory apparatus converts spacetime actions into spectra (interference patterns, or wave forms encoding phase, amplitude, and frequency) via forward Fourier transforms and in a reverse manner, constructs the three-dimensional world we perceive (spacetime); that is, spectral patterns are transformed into the objects we see out there.

Relying on Gabor's classic study on telegraphic transmissions in attempting to minimize uncertainty with message compression (1946), Pribram also posits that a domain exists that makes use of *both* spacetime *and* spectra ("Gabor-like function/domain"); that is, following Heisenberg's uncertainty principle, construction of information is at the intersection of these two dimensions which always has a degree of uncertainty. Thus this "quantum of information," or logons, are constantly being generated in any instance of interaction within spacetime. This is the field of quantum holography. A logon, as derivative, encodes both spectra *and* location in spacetime. They are "space-time constrained holograms" (Bradley, 1998: 118). They indicate how information is actualized at any particular moment; it is always in-process and never precisely definable (i.e., Heisenberg's uncertainty principle). There is, then, no possibility of one-to-one correspondence between what is "out there" and what is actualized as information. A reverse Fourier transform, then, transforms spectra back to spacetime in the form of actualities, or instantiations from the inherent potentials.

Pribram (2004: 3–4) has also noted, echoing Bohm (1983), that the eye, and all sensory receptors act as lenses; without these lenses [transducers] we would experience our world as holograms. Hence, the distinction between the actual experienced world (spacetime) and the spectral domain (frequency interference patterns). The spectral domain, composed of holograms, is a "pre-spacetime" domain of potentiality; reality is experienced in spacetime (ibid., 5). A person has a capacity to choose from an ongoing flux of spectra as to what it attends; instantiations, or actualizations in spacetime implicate a process by which a holographic pre-spacetime becomes the everyday conscious world within which we interact (ibid.).

Raymond Bradley, extending Pribram's findings to the sociological realm, has drawn from Gabor's quantum holography approach and Pribram's adaptation to develop a theory of group solidarity (1998). He also applies this to solidarity in secret societies, including terrorist cells (2010, 2011). We will return to the latter in this chapter, below. Spacetime is conceived as an energy field in flux. It is composed of a network of "bio-emotional relations" (2011: 198) whose configural informational basis is encoded and shaped by a system of control. Commonalities in groups generate a characteristic vibrational pattern (i.e., distinct frequency signature) that traverses the bounded field, much like, Bradley tells us, a key in music to which the group members can be attuned (ibid., 2006). These are encoded as "Fourier descriptors," or holograms, the spectral domain, within each member's brain. In essence, "information about the features and organization of a whole (an object or event) is spectrally encoded into the oscillations of energy as an interference pattern and is distributed to all points and locations throughout the field" (ibid., 212).[2] The oscillating waves of the group organization are said to meet with emitting oscillations from individuals in creating interference patterns (spectra) that reflect the configural nature of the bounded group. This produces the ongoing nature of group organization and provides the "glue" that is the basis of solidarity. That is, members are literally in tune and produce harmonic resonance. Drawing from Gabor (1946), Bradley then argues that this by itself lends itself to holographic

2. Bradley, is one step removed from following this implication to the second significant quantum holography approach theorized by Laszlo which states this information is in fact stored external to the brain.

determinism, but this is tempered by the actualization of ongoing snapshots suggested by the notion of logons. In other words, this produces an instantiation of the interaction between the individual and the group at any moment, the basis of which, following Heisenberg's principle is inherently indeterminate.

2. Laszlo and the "In-Formation" Field

A second approach in quantum holography has been initiated by Laszlo (1995, 1997, 2001, 2007). It has been integrated further into the notion of phase conjugation as a basis of information construction, storage and transmission (Marcer and Schempp, 1998, 1999; Mitchell, 2008; Mitchell and Staretz, 2011). Laszlo has developed the notion of an "in-formation" field within which everything is immersed. It is based on the notion of a "zero point field" (ZPF), which argues that space is not empty but a field of energy fluctuations (Puthoff, 1989a, 1989b). Recent theorizing suggests it also has compatibilities with the notion of "dark matter" and "dark energy." It is within this space, a frequency domain reflecting vibratory energy states characteristic of all entities, that all information of whatever happens is stored and is accessible nonlocally by the appropriate rays of attention of human beings.[3] In this view, memories are stored as wave interference patterns within this field. Information, then, is retrieved from this "in-formational" field (Laszlo, 2007). Laszlo suggests that it should be treated as the fifth field along with electromagnetism, gravity, and the strong and weak force of atoms. Further theorizing in this direction that makes use of the notion of resonance and phase conjugation has been conducted by Peter Marcer, Walter Schempp, and Edgar Mitchell. Thus, this approach differs with the first approach of Gabor, Pribram, and Bradley who suggest, ultimately, memory storage takes place within the brain. It argues, instead, that memory storage is encoded nonlocally in the in-formation field and is accessible by way of phase conjugation dynamics. In this view, the brain is a receiver and decoder of information from the in-formation field. It is a "transducer." It accesses and adds to this information as it actively interacts with this field. Being in tune with this holographic information is equated with harmonic resonance. In other words, when we are in tune we create channels of information flow between the emitting and receiving entity. Ultimately, however, we are all interconnected via the Akashic Field (zero point field). Laszlo's work is suggestive of a recon-

3. We will argue in the next chapter, they represent a unique signature wave.

nection between the physical and the mystical, especially of Eastern religion and philosophy.[4] Unfortunately, classical physics has atomized what is an inherently unified subject. Independently, over one hundred years ago, Bergson ([1896]2002) in *Matter and Memory* had already suggested this idea of memory storage. We will return to a more detailed examination of this approach in the section below entitled "phase conjugation."

3. Bohm: The Holofield and Holomovement

A third approach in quantum holography has been developed by Bohm (1983), who argued that the cosmos is made up of the (virtual) implicate and the (instantiated) explicate orders, guided by a "pilot wave" that is the carrier of all information.[5] This followed de Broglie's (1924) formulation, initially generating little interest, that the wave is not only a mathematical construction but has a material form, that is, it is physical. Bohm extends this to indicate that it carries information in holographic form. The totality of these waves make up the "implicate order," the virtual realm of potentialities. He suggests a fundamental interconnectedness of the cosmos ("undivided wholeness"). All entities are fundamentally enfolded in the implicate order as holograms, or interference patterns. The movement within this field of flux, between the virtual, unconscious implicate order and the actualized explicate order he refers to as "holomovement," and the actualities we experience as "relatively autonomous or independent sub-totalities of movement" (1983: 178, 189). This underlying order, he (ibid., 151) informs us, is " 'carried' not only in electromagnetic waves but also in other ways (by electron beams, sound, and in other countless forms of movement)." Further, quantum waves carry this information. He suggests these are better seen as a "quantum potential" ("Q") which carries all information in an entity's environment (Bohm and Peat, 1987: 90). It is the information contained in Q that will specify how a quantum process actualizes the real (ibid., 93). However, for Bohm, it is not necessary to make use of the notion of the collapse of the wave function (ibid., 96).

The "explicate order" is how the implicate order manifests itself in the here and now. Moreover, he promulgates the notion of "hidden variables"; that is, a primal determinate of the actualization from the otherwise cloud of possi-

---

4. The historical study by Marin (2009) provided evidence that many of the key founders of quantum theory were in touch with Eastern philosophy.

5. We will argue in Chapter 4 that some form of pilot wave could be the basis of Gibson's (1979) notion of "affordances," or virtual possibilities, to which our theorized subject diagramed as schema QD actively or inactively responds producing actualizations or instantiations from the possible.

bilities is a "quantum potential." In this view, all is intricately interconnected in one holistic universe much like the photoelectric recording plate of the "two-slit" experiments where interference patterns are recorded. Activate any part, and the whole in three dimensions will materialize. The implicate order enfolds all these interference patters, much like the recording surface in these two-slit experiments. For him (1983: 177), "the actual order itself which has thus been recorded is in the complex movement of electromagnetic fields, in the form of light waves. Such movement of light waves is present everywhere and in principle enfolds the entire universe of space (and time) in each region ..." In other words, it is in the form of light waves that information is constantly being transmitted. From such an order, may be "abstracted relatively autonomous or independent sub-totalities of movement." These are the entities we see as actualities, with degrees of order. They include memory traces. He (ibid., 207) cites favorably the work of Whitehead's (1925, 1929) idea of "actual occasions," and Leibniz's (1991) idea of "monads," each of which "mirrors" the whole in their idiosyncratic way. The primordial state of matter and energy is thus in this grand "holomovement," a perpetual enfolding and unfolding flux. For Bohm information, following Pribram's research, is not primarily stored in brain cells, but rather information is enfolded over the whole brain (ibid., 198, 208). It operates holographically. Thus, "when the 'holographic' record in the brain is suitably activated, the response is to create pattern of nervous energy constituting a partial experience similar to that which produced the 'hologram' in the first place" (ibid., 198). He does acknowledge, however, that consciousness is also more than this and "involves awareness, attention, perception, acts of understanding, and perhaps yet more" (ibid.). We pick up on this idea in the next chapter in our schema QD which is an initial approximation to a more comprehensive, inter- and intra-subjectively constituted agent.

4. Quantum Holography Principle: Black Holes, Information, and Boundaries

The fourth significant development[6] is from cosmology theory in understanding what happens when an object falls into a black hole. 't Hooft (1993) and Susskind (1995), independently, developed the so-called holographic principle which argues all information of objects that fall in the black hole and dis-

6. There are other approaches. String theory, for example, has recently suggested that the cosmos is better conceptualized in terms of two-dimensional surfaces called "brains" which are akin to the recording surface in the "double-slit" experiments.

integrate is stored on the event horizon, the boundary surface of a black hole. Bekenstein (2002) specified the geometric and mathematical limits of information that can be stored on a boundary surface enclosing a volume.[7] Maldacena (1998, 2005) provided a definitive mathematical proof of this logic.[8] Bousso (1999, 2002) extended this logic to any surface area, including boundary surfaces of volumes (i.e., "bulk," or spacetime patterns). The fourth line of development argued, in short, that any enclosed volume has all information of that three-space dimensional interior encoded in two-dimensions on its boundary surface, much like recordings in the double-slit experiments. It can be likened to Plato's classic parable of shadows, reflected on a cave wall by a fire, which are perceived as if reality. This line of thought elevated the importance of information to that of a more basic law of physics (Wheeler, 1990; Smolin, 2001; Bekenstein, 2003), including positing that information is primary and other laws derivative (Verlinde, 2010). Susskind (2009: 294) from revelations developed in discussion with 't Hooft suggested an example of this relationship between the bulk and enclosing boundary surface. He asks us to consider the room you are working in as a bounded space, the walls indeed encoding all information of the interior happenings. This gives new meaning to: "if these walls could only speak"—they do! Similarly, the classroom we teach in, the work place, the housing unit we live in—all can be conceptualized as a bounded region. Contemporary criminology and sociology often makes use of the word "boundary" rather liberally without precise understanding. The holographic principle suggests there exists a multitude of nested and overlapping boundary regions within which (the bulk) a social reality is an ongoing construction. Each cannot be isolated from the rest due to nonlocality and the phenomena of "quantum tunneling" by which holographic information is made available immediately regardless of distance.

---

7. "Surface" should not be seen in our conventional manner as indicating something physical that, for example, we can stand on, but rather is an imaginary construction with precise mathematical properties.

8. Two recent papers just published in *Nature*, on December 10, 2013, authored by Yoshifumi Hyakutake have stirred up significant interest. The papers showed mathematical proof of the fundamental insights of the holographic principle as developed by 't Hooft (1993) and Susskind (1995). These papers further acknowledged Maldacena's (1998) work that had provided mathematical support for the holographic principle. Maldacena endorsed these findings: "It seems to be a correct computation." Susskind also has acknowledged his support. See Cowen (2013) and http://www.nature.com/news/simulations-back-up-theory-that-universe-is-a-hologram-1.14328. Physicist Craig Hogan at Fermilab, Illinois, has constructed a Holometer that, when fully functional, will attempt to provide data on the holographic principle. See: http://holometer.fnal.gov/.

These four suggestions on a quantum holographic approach are provocative; they have been subject to lively discussion and research in the disciplines of physics, cosmology, and mathematics. The social sciences, and in particular, criminology remain oblivious to these challenges of fundamental ontologies created from Newtonian principles. Fortunately, there has been some inspirational work by innovative thinkers that dare to cross boundaries. The early work of Shimony (1965) and Capek (1961, 1971) have been groundbreaking. We have already seen that Bergson's classic *Matter and Memory* in many significant ways, anticipated a holographic representation of the cosmos. Whitehead's work (1925, 1929) has been cited often as incorporating early forms of quantum principles. More recently, several social scientists have risen to the occasion (Robbin, 2000; 2006, 2010, 2012; Bradley, 1998, 2010, 2011; Wendt, 2006, 2010). Some of these developments remain implicit, with Deleuze and Guattari's work being most noteworthy awaiting more explicit theorizing. For the latter, suggestive has been the work of Delanda (2002, 2006) and Massumi (1992). We want to argue that much talk in sociological and criminological theory about boundary areas without its adequate physical description can be constructively rethought along quantum holographic principles. In some cases it may be just a specification of a viable existing theory, in other cases a theory needs to be modified, in yet others, replaced as being too rigidly committed to the classical-materialist paradigm in light of the tremendous developments in quantum and holography theory. Our proposition can be stated: space can be conceptualized as made up of nested, often intersecting holographic boundary areas within which (volume, bulk) three-dimensional happenings are encoded as information in two-dimensional form on the boundary area. Thus criminologists who try and localize a boundary, be it subcultural, group, work, family, state, community, niche, geographic, etc. are asked to reconsider what they are assuming about the linguistic expression, "boundary." Our approach suggests a physicality to this term based on a process-informational paradigm. We next turn to suggesting how communication can take place within these spaces following our alternative paradigm.

Phase Conjugation: Communication Channels

In the previous chapter we have argued there is a major difference between the two paradigms described in conceptualizing the processor of information. The process-informational paradigm has formulated a novel conceptualization as to how information is constructed and transmitted. This is an alternative to the correspondence theory of classical thinking on the nature of "seeing" reality.

The notion of "phase conjugation" suggests that interference patterns are created between an emitting entity (object) and the receiving entity (percipient). It is in the percipient's unique signature wave interfering with the incoming wave ("object bearing wave") that produces an external image, "out there." Initial work of David Pepper (1986) in non-linear optics described phase conjugation, or resonance. A wave is characterized by its amplitude (height), wavelength, phase, and frequency (Pierce, 1967: 126–127; McTaggart, 2008: 25–26). One complete cycle, measured from crest to crest, is a wavelength. Frequency is measured in terms of the number of these completed cycles (or wavelengths) per second. Electromagnetic radiation such as light and radio waves travel at 186,000 miles per second. Frequency modulation means a shift in frequency. A phase is any particular moment along the wave. Waves that are in phase, or interference, mean that they are aligned with each other. This produces resonance and a distinct interference pattern; the two vibrate together in being in phase. Laszlo (2007: 70–71) offers a parable of ships at sea generating wakes that, when crossed, create interference patterns. These interference patterns reflect information about the respective ships. When more ship's wakes cross, they modulate the given interference patterns creating even more complex patterns, yet keeping the information of each within the pattern generated. When two waves interfere and where their sine waves are congruent we have resonance, an amplification (e.g., the signal is much more powerful);[9] where the two sine waves are incongruent we do not have resonance. Consider a tuning fork used to create resonance with musical instruments by tuning to the same frequency. We are familiar, too, with how glass can be shattered when a sufficient frequency is established. Where two waves meet (coincide) and one is at the low point, the other at the high point in their amplitude, we see a tendency for cancellation. Resonance, or positive interference patterns, is also called phase conjugation. Pepper's initiative is an early analysis of phase conjugation as it applies to optic research. Phase conjugation was further developed by Schempp (1992), Marcer and Schempp (1997, 1999), Mitchell (2008), and Mitchell and Staretz (2011). It has been reformulated as Pcar, or phase conjugate adaptive resonance. It is this process that provides the physicality to holographic constructions and the transmission of information. It provides the wherewithal of channels of communication.

---

9. We see examples of this in rogue waves found in oceans that have been known to create major damage on unsuspecting ships, even capsizing them.

All entities, we have argued, vibrate with distinct frequencies. They both absorb and emit electromagnetic radiation.[10] These waves also carry information, encoded holographically. They are in essence interference patterns. We live in a vast (spatial) frequency domain accessible to the person via a two-way phase conjugate adaptive resonance.[11] Bergson (2002), we previously argued, anticipated this line of thought in his image theory of the cosmos, but did not have an adequate physics to support it. We find it in quantum holography. Bergson envisioned an "object" emitting wave frequencies that are received by the brain (the "living image," or percipient) which selects according to pragmatic concerns, and sends back this selected information to the "object" which produces an effect, a particular reduced image. We experience the image where it is, "out there." Snap your finger, as an often provided example, and the perception is experienced "out there." For Bergson, consciousness itself is also experienced out there, *in* the object, not in the brain. An "object," in this view can also be any encounter that we experience. In law it can be conceived as a situation of controversy needing definition so that it can be conceptualized as having "legal standing." In other words, some controversy appears before the judge, who must construct its legal meaning in law as a "case." The abstraction process produces a particular rendition of reality which is summarized in

10. Suggested, too, is that all five of our primary senses—sound, taste, smell, touch, vision—operate under the same principle. The key sense organs being: ears, skin, tongue, nose, eyes. They are referred to as frequency transducers in the literature of quantum holography. Recent research in quantum holography also shows that the heart plays a central role (Bradley, 2007, 2009; McCraty et al., 2004a, 2004b, 2009).

11. An alternative to Pcar, the "transactional interpretation," with compatibilities but also differences, has been developed by Cramer (1986, 2001, 2006a, 2006b; see also Elitzur and Dolev, 2006). Cramer (ibid., 22) develops a four step model: the first stage argues all entities are the source of quantum waves (Ψ) that spread outward in a forward time in all directions ("offer wave"); the second stage is where some "absorber" takes action on this wave, re-emitting a time reversal "confirmation wave," a "complex conjugate" of Ψ, or Ψ*, which "travels in the reverse time direction and arrives back to the source at precisely the instance of emission with an amplitude of ΨΨ*" (ibid., 22); the third stage is a "stochastic choice" made by the source of the initial emitting wave which selects one from the many possible actualizations based on the strength of the ΨΨ* wave; and the fourth stage is "repetition to completion"—"reinforcing the selected transaction repeatedly until the conserved quantities are transferred and the potential quantum event becomes real." Instantiations, he informs us, do not take place in the brain, but "out there" at the source of the emitting waves of entities. This comports with Bergson's notion of how a perception is instantiated. Our difference with this approach is that it eliminates consciousness, or active agency, whereas in our model (see Chapter 4), the agent is the source of a distinct signature wave reflecting its existential state at any moment, which effects the perception image that eventually is instantiated as the real.

court decisions as "the facts of the case." These facts of the case are always summarized at the beginning of written opinions by appellate courts, all the way up to the U.S. Supreme Court. Of course, the dissenting judges may construct the "facts" in a markedly different way. The "easy case" in law, often said to be over 90% of the cases before it, are quickly subsumed under a particular principle of law ("subsumption," Lefebvre, 2008) and thereafter, in a linear way, or deductive logic (syllogistic reasoning), argued toward a conclusion in law. The "hard case" is where things seem a bit more unclear (i.e., cannot be easily subsumed as a typical case, subsumable to some legal principle or rule) and demands an active thought process. We shall return to the hard cases in more detail in Chapter 5.

Karl Pribram (1971, 1991, 2004), influenced by the work of Gabor and the Fourier transforms, was to revolutionize brain research in showing that it was not primarily electrochemical processes that account for consciousness, but phase interference patterns. McTaggart (2008: 84) well summarizes the process: "our brain primarily talks to itself and to the rest of the body not with words or images or even bits or chemical impulses, but in the language of wave interference: the language of phase, amplitude and frequency—the 'spectral domain.' We perceive an object by 'resonating' with it, getting 'in sync' with it. To know the world is literally to be on its wavelength." Information is also transmitted nonlocally; that is, it is instantly communicated to all parts of the body. We will return to more detailed analysis in Chapter 4 on agency.

Walter Schempp (1992) a German mathematician, influenced by the work of Gabor, was to arrive at a quantum holographic principle of brain functioning (see also an accessible explication by McTaggart, 2008: 88–96).[12] Schempp had earlier demonstrated how radar worked by a holographic principle. Some researchers have shown how dolphins and bats operate on similar principles. Schempp extended this to understanding brain MRIs (magnetic resonant imaging). The principle of the MRI was to make protons in the brain spin more actively and when they slow down, produce radiation in the form of frequencies. Information is holographically encoded in this radiation that can then be developed, by the use of the Fourier transforms, into a three-dimensional visual image by the MRI. He had also provided further evidence that the zero point field (ZPF) was indeed a storehouse of vast information, stored holographically that could be reconstructed in three-dimensional visual forms

12. McTaggart's (2008) book includes perceptive interviews with key developers and their relationships with other key thinkers and provides accessible summaries of their work. We are, in the following paragraph, indebted to her clear exposition and biographic notes.

via Fourier transforms. Then, coming across and communicating with a former student of Gabor, Peter Marcer, a "Brit," who was working on possible connections of wave theory to the dynamics of the brain (McTaggart, 91), the two went on to explain how the brain is in fact a frequency analyzer, a frequency transducer, operating by way of quantum holographic principles. An incoming wave reflecting potential forms of expression, or $|w_i>$, is decoded and sent back to its source constituting it as one image out of the possible. Said in a cruder way: the image comes in as a wave-form and returns as particle-like. These holographic transducers are also emitters of waves ("reference waves," or non-object bearing waves) reflecting the percipient's total being. The brain's cortex region and the surfaces of neurons are hologram planes (Marcer and Schempp, 1997: 521–22, 524) much like the photographic plate in the dual-slit experiments. In other words, emissions of the body are coherent quantum vibrations. Further research has shown indeed how biophotons are produced by the brain, sensory apparatus, and other parts of the body (Gariaev *et al.*, 2001, 2011; Popp, 2000, 2012). This includes the heart (McCraty *et al.*, 2004a, 2004b, 2009; Bradley, 2007, 2009; Bradley *et al.*, 2011) as the largest emitter of radiation.[13] Marcer and Schempp (1997: 522), following Pribram, argue it is the tree-like endings of nerves (i.e., dendrites) that are the "sources of holographic signals of both object bearing and non-object bearing [reference] beams for the holography taking place."

The object bearing beam, which is not to be conceived as merely a photograph, is what carries the image of the object; the non-object bearing beam is the wave, emanating from the percipient (the actual receiving person) which will interfere with the object bearing wave in a return trip along the same pathway as the incoming wave, producing the interference pattern (hologram) and projected "out there" as 3-dimensional reality. Moreover, a reference wave "may directly encode and/or transmit an arbitrarily complex set of object images as a single holographic interference pattern ..." (Marcer and Schempp, 1998: 242). Thus "each encoded analog holographic interference pattern contains geometric information concerning the incremental motion of the object images relative to one another" (ibid.). Said in another way, a vibratory entity (or encounter, situation, or moment where a complex set of waves are being emitted), can combine with the reference beam of the percipient producing one composite hologram that preserves the complexity existing in space-time, en-

13. Evidence is provided to suggest that the heart first receives object bearing waves activating not only emotionality, but also sending this vibrational pattern to the brain and the sensory apparatus from which the outgoing signature wave is created. In this view, emotional responses are experienced *prior* to the percipient's consciousness of the outside event.

coding it in an analog holographic interference pattern as information. The encoded wave of the percipient is transmitted back to the emitting entity or situation along the same pathway which produces resonance, and at its target, a particular perception. Mitchell (2008: 3) has stated it in another way: "the information carried by a quantum hologram encodes the complete event history of the object with respect to its three dimensional environment. It evolves over time to provide an encoded non-local record of the 'experience' of the object in the four dimensional space/time of the object as to its journey in space/time and the quantum states visited." Consider, again, for a moment, Laszlo's parable of ships at sea producing wakes that combine. Marcer's and Schempp's subsequent work supported Pribram's notion on how the brain operates quantum holographically. Phase conjugation, then, underlies the often referred to notion in philosophy of a "stream of consciousness," here depicted very much like a sequence of movie frames (Marcer and Schempp, 1997: 530; 1998: 242–244).

After Marcer and Schempp wrote several papers (1997, 1998, 1999) on this topic, Marcer was to meet Mitchell, a former Apollo astronaut. He was presenting a paper at a conference. According to McTaggart's interviews (2008: 91), Marcer and Mitchell were to have several meetings together discussing their mutual interests, and seeing compatibilities, they decided to work together. Schempp, too, was to ongoingly share information with Pribram about their findings (ibid.). Mitchell was to do further refinements gained by their joint information sharing (2008; Mitchell and Staretz, 2011; see, again, McTaggart: 2008). Together they confirmed that objects were vibratory entities that existed in the zero point field: that we didn't directly perceive them, that their informational basis is spectrally encoded, that this information is accessible via phase conjugation and Fourier transforms, and, finally, that nonlocality assures instant connectedness within the body and with the zero point field.

Mitchell's (2001, 2008; Mitchell and Staretz, 2011) significant contribution to the quantum holographic model being developed by Marcer and Schempp was to refine the nature of phase conjugation ("phase conjugate adaptive response," or Pcar) and to tie it more directly to the connection between each person's (percipient's) unique signature wave in accessing information in the ZPF or the in-formation field. The ZPF is the storage field, then, of information, not the brain as in Pribram's approach in quantum holography. Mitchell explains that "attention" is a necessary component for phase conjugation to take place. Finally, he shows how during the resonance phase, that is, once phase conjugation has taken place in the form of a standing wave, there is an "evaluation" process by which the percipient accesses its memory (prior experiences and beliefs) in order to compare this information with the holographic image. From this, "intent" is established in relation to the object (2011: 14). In other words, this is a precursor for conscious motor action. Quantum theory's notion of nonlocality assures that every cell of the percipient is in resonance during phase conjugation. We shall return to Mitchell's model for its relevance to agency (Chapter 4) and to the development of a transformative justice (Chapter 6).

Laszlo (1995, 2007) has explained how the zero point field is an "information" field. Coming full circle to Bergson's work of 1896, memory storage, he argued, is not in the confines of the brain as in traditional research, but exists spectrally encoded in the in-formation field. His parable was of the sea modulated by the interference patterns of ships plying the waters traces of which remain encoded as informational bearing memories. The sea was likened to the ZPF. This information or hologram would remain indefinite save for the canceling properties of shorelines, wind, and effects of gravity (ibid., 71). Consistent, too, with Bergson's analogy with a telephone switchboard, the brain is a frequency analyzer, a frequency transducer that accesses the information field with its own unique signature wave emissions producing three-dimensional images we "see" as reality. Moreover, one version of the holographic principle ('t Hooft, Susskind, Bekenstein, Bousso, Maldacena), as we have previously seen, argues that this information is further encoded in two-dimensional form on boundaries that surround a volume. We could also conceptualize these bounded areas as noospheres (Deleuze, 1989) as well as "ecological niches" (Gibson, 1979), their unique coordinating sign (symbolic) systems as noosigns, and a person's unique accessible information (by their emitting signature wave) as her/his quantum lexical structure. We will further develop the dynamics of agency and the quantum lexicon in the following chapter.

Let us provide an example of phase conjugate adaptive response. We come across a situation, a state of affairs, a spatiotemporal happening, which emits vibrations and wave frequencies, the basis of the object bearing beam. To simplify, take what we will eventually construct as an apple. This entity emits wave frequencies reflecting its constituent elements. The percipient's transducers receive this information while emitting its own distinct signature reference beam. The combined holographic interference beam (phase conjugation) is reflected back along the same pathway to the source, where it is seen as an image. This image then returns to the percipient and becomes the basis of a motor response. Images are constructed "out there," not in the brain. Consciousness, too, exists *in* the images created, "out there." In this view, all with which we interact retains a trace of ourselves, stored as an analog holographic interference pattern. Bergson's remarkable insights from 1896 have found verification. It is here also where mystics and Eastern philosophy have found compatibility in arguing that there remains something permanent of each and everyone one of us and our activities in the zero point field (ZPF).

Let us provide a more concrete example. Consider juveniles hanging out on a street corner. There is much movement in this locale. The juvenile receives these frequency waves in complex patterns which are decoded by frequency transducers of the body. These transducers are always creating their own frequency waves, signature biophotons, reflective of the totality of the person's being and becoming. The interference pattern created by the incoming and outgoing frequency pattern produces a hologram, a carrier of information, that is projected out there producing the image of the street corner setting, to which s/he may respond through motor action. Some juveniles may indeed evaluate the situation as threatening deserving a violent response, but yet others might construct the situation quite differently. Creating it one way rather than another is equivalent to saying the wave function of potentialities collapsed one way or the other. That is, the former accesses memory images from the ZPF that may be heavily populated by rationalizing discourses which, entering resonance with the juvenile, change attention to intention; the latter activates "templates for action" that include modal forms of violent responses in that milieu. We have also seen earlier that since particular instances of reality constructions involve the construction of spacetime constrained holograms (logons) that are somewhat indeterminate, that a slightly modified perception image created might not necessarily lead to a violent intent and a violent response. We shall have more to say in Chapter 5 concerning application of quantum holography to criminology. Creating an event, an "actual occasion" to use Whitehead's (1925, 1929) language, is the process of creating a particular reality ("concresance"). In the next chapter we will also demonstrate that a bona fide conception of agency would show how this process is inter- and intra-subjectively constructed in producing a social reality.

Exemplary Initial Applications of QH from the Social Sciences

Two exemplary applications of quantum holography stand out in the social sciences. Alexander Wendt provides insights by applying quantum holographic principles to theorizing international relations. Bradley applies it to group solidarity, intuition, and terrorist cell research. They have also extended on the development of a quantum consciousness to include holography theory.

1. Alexander Wendt: International Relations and Quantum Holography

Wendt's grand approach includes an interdisciplinary integration. He embraces classic philosophers Spinoza and Leibniz and more recent scholars such as Whitehead with their "panpsychism" approach; that is, all entities are seen as possessing degrees of consciousness. In Wendt's read of panpsychism consciousness does not collapse the wave function but happens during the process of the collapse (ibid., 286). Human beings, the State, international relations, and all else, would ultimately be wave-particle like entities. The brain, too, operates quantum holographically, following the work of Pribram. He also follows the work of Peter Marcer and Walter Schempp who describe the process of phase conjugation as responsible for how perceptions are manifested from the virtual. He draws from Gibson's ecological approach indicating that any field offers "affordances," or in his view, a wave function of delineated possibilities that are essential ingredients in subsequent perceptions developed by subjects. He also draws from Nakagomi's (2003, 2006) work on "quantum monadology," a revisionist version of the work of Leibniz (1991) concerning monads. Each monad, in this view, conceptualizes the world idiosyncratically and their interior states are composed of holographic models of this world (Wendt, 2010: 290). Finally, he draws from language literature in showing that language too can be better understood as quantum holographically constituted. Texts are inherently waves of potentiality, a virtual realm; particular meaning is constructed via interaction between the reader and the text. The collapse of the wave function produces a real from the virtual; a particular nuanced meaning associated with a particular text. He (ibid., 293) concludes "if language is a quantum phenomenon, then social objects like the state and international system that are constituted by language will be quantum mechanical too." They are holographic projections.

He concludes his provocative and highly illuminating examination by "bringing individuals back in"; that is, unlike classical physics where there is little place for consciousness (recall, all is ultimately determined), his approach insists on the viability, the centrality of consciousness in how international relations are constituted as "objective" entities. In his view (ibid., 296), human beings are not necessarily passive observers but actively engage the virtual with its potentials, and all is nonlocally interconnected. Each person is a member of a "social wave function" that represents the State. State leader's decisions produce a collapse of this social wave function that instantiates a particular real from the possibilities embedded in the virtual (ibid., 297). He further distinguished between "active monads" and "passive monads." Each human being is entangled within a plurality of social wave functions, and only in some is s/he in an active mode. In the active mode their attention is more directly in tune with the particular potentials that exist. In a passive mode, they are not, and more likely more attentive to other social wave functions. Wendt (ibid., 300) defines the "passive monads as being entangled quantum mechanically, at the level of the implicate order, which gives them the capacity to become subject (active) in their joint holographic reality if they choose." He also identifies a third form, those who are merely objects; that is, they are not entangled with the system and thus do not have the direct capacity to become active monads. It is only through an educational process ("acquiring relevant knowledge") that entities can become first passive monads, then potentially, active monads. He (ibid., 302) is quick to also remind us that "consciousness is what happens when wave functions collapse," and this process always takes place amongst quantum entangled persons; hence, there exists an inherent I-you relationship rather than a self-enclosed "I." "I" exist in you, you exist in my "I" (citing Kolak, 2004). Not fully developed here, but suggestive is an ethics. To this we will return in Chapter 4 and 6.

In the closing pages, Wendt considers the wherewithal of invariant structures that are harbingers of potentialities. Human beings find themselves quantumly entangled in a plurality of social wave functions representing these structures, an implicate order. In our furtherance of his initiative, we shall return to Deleuze's notion of assemblages as a bona fide direction that remains true to quantum holographic principles. We shall also integrate the idea of relatively autonomous bounded regions which are structures consistent with these ideas. Wendt concludes with a call for a process ontology that integrates the person with structures in understanding how the virtual possibilities are rendered real by the collapse of the wave function. For him, it is an individual "choice" that collapses the wave functions.

2. Raymond Bradley and Group Solidarity

A second exemplary application of quantum holography in the social sciences is the work of Raymond Bradley. Bradley, a sociologist, has provided several articles in applying quantum holographic theory to sociology. Unfortunately, it has not taken off in the rigid, dogmatic field of academic sociology. Two areas are especially noteworthy: his work on solidarity, and his work on secret social groups, including terrorist cells. He has also applied this analysis to the nature of intuition in business affairs.

In an early article, the nature of social groups was examined by Bradley (1998) applying the conceptual framework of quantum holographic theory.

He distances himself from Laszlo's early version of classical holographic theory, arguing it is a form of holographic determinism, and instead, embraces a quantum holographic approach rooted in Gabor's (1946) classic work on the nature of quantum information, the basic unit being a logon. The latter approach is further traced to the work of Schrödinger and particularly Heisenberg's uncertainty principle.

In Bradley's conceptual scheme, there are two major determinants of momentary instantiations of information, flux and control, which, when recurring in collaborative communication, are the basis of group solidarity. These can be placed on two perpendicularly intersecting axes. Flux, the horizontal, represents potential energy, passions, flows, bio-emotional energy. Reconceptualized in Deleuze and Guattari's frame, this is the "plane of immanence," a preconfigured space of pure energy and movement. Control, the vertical, represents "the degree to which individuals are interconnected by a transitively ordered network of relations—exercised at that location" (Pribram and Bradley, 1998: 293). It is a system of hierarchical power relations (Bradley, 1998: 128), a "system of social constraints" (ibid., 130). Referencing Piaget's classic work (1995b), this constraint dimension includes language, "a common system of signs and references" (ibid., 148) as well as nonverbal communication (Bradley, 1998: 130) (e.g., body gestures, postures, etc.). Suggesting social collectives share group boundaries, Pribram and Bradley (ibid., 292; Bradley, 1998: 128, 130; Bradley, 2010: 138) see these patterned interrelationships located in a "dense web of mutual relations of positive affect" as a "field." This field, he continuous (2010: 138; 140–42), "operates to unify and activate affective attachments, arousing each members' emotions and thus their potential energy—their propensity for expending bio-emotional energy in action." Within each bounded area various configurations of control and flux produce logons, or quanta of information. It is these that are mobilized and reenergized in use in producing the regularities, stabilities and solidarity experienced in social collectives. In other words, agents find themselves in these fields reconstructing logons in more typical manners associated with flux and control in constructing social reality. They remain literally in tune with each other. It should be stressed that these are not rigid due to the Heisenberg principle underlying the construction of logons. Thus a degree of uncertainty prevails. Conceptualized in terms of dynamic systems theory (chaos theory), identified is a phase space with constituted logons acting as attractors (ibid., 295). Moreover, Gabor suggests logons are not discrete but rather overlap, they produce an "overlap with the future" (1946: 437). That is, information encoded in one logon is also shared with an adjoining logon. What this suggests is that as agents construct logons they are always already in the anticipatory mode as to information that will follow. This can also be the basis of intuitive responses within a social collective. In sum, the ongoing exchange and reification of logons acts as a social communication system producing a common ethos, understanding, cultural order, solidarity, and an enduring subculture.

Bradley specifically elaborates on the nature of purposive, intentional action. Citing the works of Freud, Piaget, and Pribram, he (ibid., 139) argues "thoughts are implicit acts"; in other words, the perception image we create always already has implicit a plan of action. To justify this, he draws from Piaget's (1995a, 1995b) analysis of cooperative systems that take place in "the context of a common system of language, values, and social norms" (ibid.). Piaget's concept of "groupement," a homeostatic process, in which sensorimotor, cooperative, coordinated activity becomes normative, provides the grounds for the transformation of thoughts/perception images, with their implicit "virtual actions," into purposive action (ibid.). Furthermore, Piaget's insights are reconceptualized: Bradley's idea of flux is now accommodative to information flow amongst individuals; control, to Piaget's "common system of signs and references" (cited in Bradley, 1998: 142a), or for Bradley, "common language values, and social norms" (ibid.). This is what produces stability, the basis of an underlying movement to equilibrium in social organizations and is the basis of circumscribed individual purposive actions. In short, purposive action is situated within the parameters of flux and control; thought/perception images that are produced within this context are implicitly virtual actions, which, within a system of control, will materialize in patterned forms. We thus have a feedback loop that produces both consistency and stability in a social group on the one hand, and the basis of a circumscribed basis of purposive action, each of which reinforces the other. Applying classical holography, we see how patterned forms are interference patterns encoding holographic information. Repeating linear forward and reverse Fourier transforms reify dominant cultural norms. In Bradley's words (ibid., 124), "spontaneous or purposeful behavior (novelty or intentionality) is not possible in classical holography because everything is 'encased' in a predetermined order." However, applying quantum holography of Gabor (1946), we see that ongoing constructions of purposive action are based on thoughts/perception images that have, following an adoption of Heisenbergs' uncertainty principle, degrees of indeterminacy; they are logons, or constrained spacetime holograms (ibid., 127). In other words, the intersection of flux and control produces ongoing logons which provide a degree of variation from any cultural or subcultural norm, thus providing the basis of nonlinear volitional behavior, change, and creativity.

In an article in collaboration with Karl Pribram (1998), "The Brain, the Me, and the I," inspired by Piaget's discussion of accommodative and assimilative principles as the basis of sensorimotor action, they indicate that two key components of the agent/self are at work, the "me" and the "I." The "me" is an ego shaped by environmental factors; the "I," monitors and evaluates the actions of the "me." The "me" is a reflection of Piaget's assimilation and hence constancy and stability; the "I," accommodation, the basis of change. Together, these two processes produce a social self. The interaction between I-me plays itself out as group members in collaborative activity find themselves within a field composed of two intersecting axes of flux and control: flux, the horizontal, concerns potential energy expenditure (i.e., people engaged in work); control, the vertical, social constraints, including language, norms, values. Flux is conceptualized as the spacetime domain ("free exchange among individuals," fluctuations within the 3-dimensional everyday spacetime world); control, the spectral domain (relatively stable, more static frequency interference patterns enfolding holographic information). The intersections of these two dimensions in everyday collaborative activity produce quanta of information, or logons which are the basis of ongoing communication, cooperation, and stability. These two dimensions bound a phase space where distinct attractors take shape; that is, where regularities can be identified. Regularities, they suggest, are not so much in the form of Piaget's "mobile equilibrium," but rather more in terms of far-from-equilibrium conditions and, we add, dissipative structures of Prigogine's (1981, 1996) dynamic systems theory approach. A stable self thus entails accommodative and assimilative processes, dynamics that include the dyad I-me immersed in a field constituted by flux and control. Informed by Piaget (1995a), it is via everyday cooperation or in Pribram and Bradley's appropriation, collaboration, within a shared system of norms, language and values that a stable social self is produced (Pribram and Bradley, 2001: 291).

In more recent works, Bradley (2010, 2011) has applied his quantum holographic approach to secret societies including terrorist cells. Bradley (2010) endeavors to "establishing a method of detecting membership in secret social groups, such as cults, criminal gangs, drug cartels, and terrorist cells." He also includes vigilante groups and covert militias (ibid., 126). Moving away from his earlier position that was dismissal of classical holographic principles of Laszlo, he now embraces both the classical and quantum holographic versions.

He applies classical holographic principles, as envisioned by Gabor's (1948) short article, to indicate how a social collective's identity attains codification and remains a constancy embedded in each agent's brain. He applies Gabor's (1946) quantum holographic oriented statement to show how ongoing interactions within the group construct logons as a basis of every momentary interaction and communication. The former is about image processing and the Fourier transform is the mechanism by which everyday spacetime reality is transformed into the spectral domain. Information is encoded nonlocally. It is accessible by all members of a social collective. These represent patterned interference patterns and resonate throughout the field as the group's distinct signature wave. Members are in tune with this wave. In other words, resonance is established producing a common understanding. Members constantly activate this field which intensifies their sense of solidarity. These interference patterns or Fourier descriptors, Bradley tells us (ibid., 143, 146), are holograms and are neurologically encoded in the brains of agents. Collectively, these holograms make up a static holographic signature wave of the group, its distinct cultural identity. It manifests in shared expressions, symbols, language and culture (ibid., 145). Retrieval of these holograms (Fourier descriptors) is via a reverse Fourier transform transforming the spectral domain into a spacetime instantiation of social reality. In short, classical holography explains "how information about the features of the group's sociocultural identity, as a relatively stable object of constancy, are spectrally encoded as a static holographic signature and communicated in the interactions in which members are involved" (ibid., 146). Social groups and organizations can then be seen as bounded spaces, a field, within which members are in tune with the distinct signature resonance frequency. Members have access to this signature frequency and through its ongoing access contribute to its amplification (ibid., 147). This feedback loop produces collective solidarity and the basis of ongoing social reality construction.

He then applies a quantum holographic approach of Gabor (1946) to indicate how every day, ongoing changing activity is codified as quanta of information (logons). Ongoing flow is cut up into distinct blocs of spacetime. Gabor, we recall, studied the minimum degree of uncertainty that can be identified before information loses its coherent basis, drew from Heisenberg's uncertainty principle in showing that two perpendicularly intersecting axis ("orthogonal coordinates," in precise language), representing energy-frequency and spacetime (spatial coordinates), produce a "spacetime constrained hologram," a logon. Each logon reflects momentary construction of information as a result of member's interactions. In recurring behavior it is a momentary "pattern of energy expenditure." These spectra, or interference patterns, follow the logic of forward and reverse Fourier transforms; the forward dimension constructs spectra or "Fourier descriptors" which represent the ongoing snapshots "encoding the set of energy and spatial frequencies that describe the group's interactional patterns at that moment" (Bradley, 2010: 147) and are nonlocally distributed throughout the field; in the reverse process, the brain decodes and converts these spectra into the three dimensional imagery (and sensations) we perceive as "reality" (ibid., 149). Thus the vertical dimension of hierarchical control produces a distinct cut of ongoing flow of energy, producing to use Deleuze's words, a "bloc of spacetime." These are quanta of information. They are logons or holograms.

These two processes, classical holography and quantum holography, work together to produce a pattern matching process whereby the group's nonlocal quantum signature frequency is ongoingly compared with the logons constructed at particular moments of social interaction within the group. Where there is pattern matching we have resonance; members are in tune. This is a basis of felt solidarity of the group. Where there is discord between the two holographic principles members are provided the basis for modifying the social collective (Bradley, 2010: 152–53).

Applying this to terrorist cells in particular, Bradley (2010, 2011) argues that massive governmental surveillance to determine its membership are grossly inefficient when compared to the possibility of applying holographic principles (both classical, Gabor, 1948, and quantum holography, Gabor, 1946) in identifying pattern matching instances taking place. In other words, terrorist groups, like all social collectives marked by boundaries, have distinct signature frequency waves (e.g., distinct resonance, or spectra). When members communicate with each other they access these spectra as they encode momentary quanta of information. Information is holographically enfolded in the energy spectra of a group's vocal activities. Ultimately, For Bradley (2011: 212–213), following Pribram's research (1991), holographic information is stored within the brains of its members as Fourier descriptors. Forward and reverse Fourier transforms account for how images are, respectively, encoded as spectra, and reconstructed as an everyday three-dimensional spacetime social reality. Members are "in tune" with the group's signature wave; non-members are not. Being in tune means they are in resonance (2011: 206). Any collective encompasses a field of oscillating energy—common frequency waves encoded with holographic information about that group—that radiates within the group. Group solidarity is where harmonic resonance exists. Theoretically, the resonance created can be identified, and so too the members (2010: 153; 2011). The first step in identifying the group and its members would be to discern and specify the group's distinct signature wave. He (2010, 2011) outlines how we could use a form of spectral analysis of vocal interactions to identify any constancy in the frequency waves generated. Once established, unknown members could be identified by matching their vocal recordings against the established unique signature wave of the terrorist cell. This would be the basis of further corroboration.[14]

14. We are aware and concerned that dissident group members can also be identified by this same process.

Assemblages, Strata, and Abstract Machines

Deleuze and Guattari (1987) offer alternative concepts to contemporary approaches rooted in the classical-materialist paradigm in how society is organized. These concepts can be augmented with integration of concepts from quantum holography. To this end we want to show how quantum holography can be applied in explicating several key concepts: assemblages, strata, and abstract machines. That is, given a cosmos of vibratory energy, a quantum field of fluctuating energy directed by the Schrödinger equations of motion, a "plane of immanence" (Deleuze and Guattari, 1987), a virtual field of possibilities, the question arises as to how instantiations can be described in their singular and interconnected (social collective) forms.

Let us be clear about the "levels" Deleuze and Guattari employ. From Delanda's (2002: 9–44) discussion on "multiplicities" as opposed to essences, we can draw three ontological levels: the intensive, which is the primordial vibratory energy field, closer to what is conceptualized as the ZPF; the plane of immanence, the loci of the virtual and from our incorporation, of the holofield composed of frequency interference patterns; and the plane of organization, the actualized real, a striated space, the more static structures we perceive as reality. We shall often combine the first two and refer to it as a holofield, in-formation field, or ZPF in this book. We shall see that the more rigid forms of organizations, "molar assemblages," are more connected with the plane of organization and come to be by a uniform repetitive collapse of the wave function. "Molecular assemblages," what chaos theorists would refer to as dissipative structures, find their natural home between the intensive and plane of immanence. Their characteristic is not repetition of the same, but a becoming-other. For Deleuze and Guattari, not essences, but "multiplicities" are the basic units of organization, standing for complex, but yet somewhat stable forms that are undergoing continuous change (see Delanda, ibid; Patton, 2000: 35–38, 42–43). Assemblages are maps of these multiplicities.

Social organizations[15] are assemblages as conceptualized by Deleuze and Guattari (1987; Delanda, 2006; Murray, 2013: 67–84). That is, building on Bradley, Laszlo, and Gabor, social organizations are constituted by characteristic frequency interference patterns embedded in the ambient holofield and further encoded on boundary surfaces which are continuously recreated in their use. Assemblages describe how patterned energy forms, geometrically encoded information (frequency interference patterns), take on more enduring form;

15. They are defined as having a "shared membership (a boundary) and collaborative interaction relative to a common purpose or goal" (Bradley, 1998: 123).

how, from this, momentary spacetime constrained holograms, or logons are constituted; how interactions instantiate percipients, and assemblages produce events or "occasions of experience"; how an underlying logic, an abstract machine produces repetitive effects; and how all this is historically sedimented in a particular form, what we see as everyday reality, our socius.

1. Assemblages

The idea of an assemblage is initially developed in Deleuze and Guattari's *A Thousand Plateaus*. It is further developed in Deleuze's (1988) revisionist analysis of Michel Foucault's work, particularly *Discipline and Punish*, *Archeology of Knowledge*, *Birth of the Clinic*, and *Madness and Civilization*. It suggests that all enduring substances (entities) or multiplicities are constituted along two perpendicularly intersecting axes: content-expression, territorialization-deterritorialization.

"Content" (sometimes referred to as the material role) stands for how bodies are oriented, "of actions and passions, an intermingling of bodies reacting to one another" (Deleuze and Guattari, 1987: 88). For example, consider the prevailing form of orientation in a traditional farm community orientation compared to an assembly-line workplace and to the contemporary cyber workplace. Consider the discovery of the prison (Foucault, 1977) as a new way of acting on bodies, the body of the prisoner. Here space and time is regimented, compartmentalized, subject to an ongoing gaze and forces of correction and normalization (e.g., extreme form of "control," specified by Bradley, 1998). "Expression" refers to both linguistic and nonlinguistic statements. For example, language can be an example of the former, body gestures, facial expressions, dress, and non-verbal statements of solidarity the latter.

"Territorialization" refers to how energy is bound, given more enduring or stable form; in other words, how spatial boundaries are created and given more enduring identity. It tends toward order, and homeostasis. "Deterritorialization" stands for the ever-present forces of destabilization, change, mutation, transformation, metamorphosis. Consider, for example, the inherent conflict in law between principles of formal rationality and substantive rationality, what Max Weber (1978) referred to as the "insoluble conflict." Weber sees these two form of rationality inherently embedded in conflict in emergent capitalist societies. Think, also, in terms of an oppositional group united by a singular agenda but undergoing continuous internal tension because of a multitude of other simultaneously distinguishing factors of its members. Only on some, or even perhaps even one, does resonance exist as a basis of a unified oppositional front (see Hardt and Negri, 2005). In the critical legal studies movement we have witnessed how internal differences produce and continue to produce various derivative perspectives (i.e., Crit-Fem, Crit-Lat, queer theory, critical race theory, etc.).

Delanda's (2006) explication of assemblages shows its applicability to various social entities: persons, social encounters, interpersonal networks, organizations, states, cities, and nations. For a *person*, these are suggested components: form of content (material role) include all internal bodily forces including passions that are behind sense impressions; form of expression include linguistic (shaping beliefs) and nonlinguistic (e.g., ideas); territorialization, habits and their repetition; deterritorialization, anything that disrupts habitual repetition (e.g., sensory deprivation, intoxication, etc.) which also includes new skill development that can increase the capacity to effect and be affected. A *social encounter*, as a form of assemblage being exemplary in Irving Goffman's work, could include: content/material role—co-presence, attention and involvement; expressive—public appearance such as facial gestures, posture; territorialization—reciprocal acknowledgement, turn taking, and assumed time constraints within a bounded space; deterritorialization—anything that undermines conversation or puts into question boundaries. He offers thoughts from Irving Goffman (1982, 1986) and how people in social encounters try and overcome moments of embarrassment and humiliation. Another example of an assemblage is *interpersonal networks*. It includes: content/material—lengthier, recurring amount of energy and time invested in maintaining common stances (properties of which are made up of frequency, strength, emotionality, reciprocity as well as density and stability); expressive—actions indicating trust, solidarity, pride, sometimes with nuanced language styles; territorialization—stories and conceptualizations that are shared, shared boundaries, conflict with non-members; deterritorialization—social mobility, reductions of rituals.

It is Bradley's work, however, that provides illumination as to how social assemblages are created as informational and communicational phenomena. Assemblages, in a quantum holography approach, can be conceived of as more stable frequency interference patterns, organized or patterned energy, that have attained more stable form, and which are continuously reconstituted in everyday use. Spacetime constrained holograms (logons), recall, are created from this background frequency and spectral domain. Assemblages, following an idea from Spinoza's *Ethics* can also be conceived in terms of their capacities to effect and be affected (Deleuze and Guattari, 1987; Delanda, 2006). They can also be conceived in term of signature threshold values which characterize their causal capacities to effect and be affected. Delanda (2006) refers to this more nonlinear notion of cause as "catalysis" (see Chapter 2). Assemblages can be mapped as "diagrams" an abstract mapping of vector forces, as possibility spaces with degrees of freedom and unique attractor states defining the resting point toward which they tend (Delanda, 2006: 30). As Bogue (2003b: 83) explains, assemblages are "virtual planes of 'lines of flight,'" or patterned energy in movement; they are "tendencies, becomings, directions of movement toward potential actualizations of various sorts." Some assemblages remain more static; these are molar assemblages. Some are more in the form of dissipative structures, coming into temporary being, dissipating, and restructuring, with sensitivity to even very small perturbations. These are molecular assemblages. Roberto Unger (2004), for example, has developed a notion of an "empowered democracy" constituted by core institutions that seem very much like molecular assemblages. So, too, Hardt and Negri's (2005) notion of the "multitude" as an alternative to static group identities.

Bradley's notion of social organization can be reconceptualized as assemblages populated by a bounded network of yet other nested assemblages at different scales. As patterns of energy, or information, they are continuously constituted in everyday instantiations (logons) (Bradley, 1998; see also Wendt, 2006, 2010). As quantum units of information, or logons, they then remain holographically encoded within the holofied or in-formation field (Laszlo, 1995, 2007) and are further inscripted in one less dimension on the boundary surfaces ('t Hooft, 1993; Susskind, 1995, 2009; Bekenstein, 2003). They become part of this overall information encoded within the particular milieu within which they are initially constructed. The repetition of a distinct signature wave marking the identity of an assemblage further reinvests and reconstitutes the distinctiveness of an assemblage (e.g., subcultural identities). Assemblages, seen as bounded spatiotemporal loci, are multilevel and only relatively autonomous as they interact with other spheres. Bradley (1988: 123) has suggested that the interaction and patterning of flux and control is how social collectives take form. "The interaction between flux and control operates as a communication system. It generates logon like units of information and transmits them throughout the collective." He continues (ibid.), "because each unit contains an image of the collective's internal organization, the communicative system *in-forms* (gives shape to) the *inter*-action among members and produces a stable organization" (emphasis in the original). To revise Bradley's conceptualization, however, his second variable in producing stable order, "control," can be reformulated by applying assemblage theory and how order arises from a primordial universal variation, or plane of immanence in molar or molecular form.

2. Strata

Deleuze and Guattari (1987) describe the cosmos as an intensive field and a plane of immanence, a quantum field of flux and flows of vibrational energy with periodic emergence of multiplicities. We reconceptualize the plane of immanence as the holofield, an in-formational field composed of frequency interference patterns and of dynamically (spacetime) constituted contextual information in the form of logons. It tends toward being populated by assemblages. The plane of immanence, for Deleuze is composed of all sorts of modulations: "sensory (visual and sound), kinetic, intensive, affective, rhythmic, tonal, and even verbal (oral and written)" (Deleuze, 1989: 43–44, 29). Following Bergson's (2002: 33) notion of the universe as a vibrational whole, a "gaseous state" of "matter-flow," Deleuze (1986: 58) has it that the cosmos is "... a world of universal variation...," a sea of images in movement, "the thing by itself, the image in itself, is a virtual image, a given 'block' of light, its perception an actualization of a portion of that block" (Bogue, 2003a: 34). Thus matter is light; it is a matter-flow where there is "no division between motion and things moving" (Bogue, 32–33; Deleuze, 1986: 34). It is composed of movement-images, the entities themselves in movement. We appropriate this to mean these are quantum holographic waves of information encoded in the ambient holofield and on boundary surfaces. The "living image" (the person) is what instantiates a real out of the virtual. We will argue that it is the subject conceptualized as schema QD (next chapter) that collapses this wave function and is in turn collapsed to give momentary form and substance to otherwise images in movement. This is via forward and reverse Fourier transforms, collapse of the wave function, and phase resonance. Assemblages are instantiated unities with varying degrees of longevity. Where assemblages interact they produce events, actual occasions of experience (Whitehead, 1925, 1929).

Assemblages are historically constituted in relatively autonomous ordered constellations, or strata. These are bounded regions of psychobiosocial activity. Strata, or more enduring aggregates, for Deleuze and Guattari (1987), are areas of more stable, resistant, and recurring configurations of forces. They code and territorialize the plane of immanence producing distinct layers of regularities. We appropriate this to mean holographic information. Strata are layers "giving form to matters ... imprisoning intensities ... [and] systems of resonance...," organized as "molar aggregates" (ibid., 40). As molar entities they follow homeostatic principles and tend toward particular attractor states. Chaos theory would suggest the primacy of point and cyclic attractors. Strata capture ongoing flows of quantum fluctuations. Any milieu is vibratory, "a block of spacetime constituted by the periodic repetition of the components" (ibid., 313). It is the "refrain," or rhythms, melodies, and resonances that is the basis of the emergence of territories and strata (ibid., 322, 348). Thus, as content is being given form, a becoming-content, so too are forms of expression, a becoming-expressive. Repetition and consistency produce forms of expression or semiotic systems (ibid., 334). We are at the beginnings of the emergence of a plane of consistency, and in a more enduring form, a plane of organization. It is on this plane of consistency or organization that matter clumps, or stratifies in distinct layers (ibid., 337).

We now see how social collectives are strata formations either more molar or molecular in form. An example of the former being a rigid bureaucracy, the latter a jazz ensemble. Forms of strata, therefore, mold the flux-flow of matter energy in the ZPF. The interactions produce both spectra (frequency interference patterns), or holograms, and contextually and situationally based logons, or spacetime constrained holograms (Bradley, 1998). Enclosed information is accessed by members of a particular stratum by a common signature wave and phase conjugate resonance (Pcar), and are continuously reconstituted in their use. This produces the stability we find in social collectives. The cultural framework of a social assemblage/strata is therefore regulatory and normative (ibid., 132). The interaction between strata (social collectives/assemblage) and flux produces logons or units of information reflecting internal organization and energy distribution (Bradley, 1998: 131). Such factors as "fear, hatred, or jealousy" will decrease "harmonic resonance" reducing potential metamorphic outcomes and increasing homeostatic tendency, whereas "admiration, awe, or love" increases harmonic resonance producing more energized activity within the social collective (ibid., 138–39). Logons, as quantum information, are themselves never precisely determinable because (1) as "space-time constrained holograms" they reflect Heisenberg's uncertainty principle, and (2) by the fact that logons, as sinusoids, analog/spectral in form, overlap with other sinusoids and are therefore always composed of other logons. As a consequence, each anticipates information, what Gabor (1946: 437) called "overlap with the future." With Deleuze and Guattari we add that because of the ubiquitous forces of deterritorialization, logons are always subject to possible revision and reconstitution. On the latter point, Bradley suggests fidelity in communication necessitates "constant exogenous regulation of the signal-producing device" (ibid., 143). This means that logons need constant ordering in order to remain relatively stable and to be efficacious. Bradley also suggests that it is the active nature of information processing by the brain that assures further nondeterministic outcomes (ibid., 143). He does not offer a sustained, in-depth statement of agency itself (beyond a brief reference to the work of Piaget), a point to which we will return in the next chapter. Quantum holography, in short, contrasts with any form of reductionism that reduces all logons to static structures and determinism.

Strata can also be further understood in terms of "seeing and saying." Strata are, therefore, characterized by two dimensions: "visibilities" and "sayabilities" (Deleuze, 1988: 47), content and expression. "Visibilities" are what can be extracted and perceived from all of the potentialities that exist on the plane of immanence. In applying this to Foucault's (1977) analysis of the origins of prisons, the discovery of the prison-form and a new way of acting on bodies (panopticism), produced new visibilities and new matter (prisoners). They are "forms of luminosity which are created by light itself and allow a thing or object to exist only in a flash, sparkle, or shimmer" (ibid., 52). In other words, these are quantum fluctuations, and momentary instantiations of the virtual when the wave function collapses. Deleuze hastens to say that they are not only defined by seeing, but can be "multisensorial complexes"—"complexes of actions and passions, actions and reactions" (ibid., 59). Each historically formed stratum circumscribes forms of visibility (ibid.).[16] The modern state, for example, discovers prisons which makes crime visible (Foucault, 1977). It is a "system of light" (Deleuze, 1988: 32) governed by panopticism, a method of surveillance, control, and normalization. Each historical formation creates "places of visibilities" such as prisons, hospitals, the workplace, schools, etc. It also provides the possibilities for various discursive subject-positions—identities, we conceptualize as the basis for expressions of an "I" in discourse, constituted by schema QD—in interaction with historically defined "technologies of self" (Foucault, 1994). The latter refers to how we learn to perceive ourselves, that both actualize these potential visibilities and are actualized by them (more on this, next chapter).

"Sayabilities" are forms of expression, such as linguistically constituted statements. Languages, or "regimes of signs" allow certain narrative constructions. These regimes of signs are the noosigns found within any noosphere. Lacan (1991), for example, has offered the idea of the four structured discourses (master, university, hysteric, analyst), each of which leads to a different way of embodying desire and constructing social reality. Paradigms in criminology (consensus v. conflict; consensus v. critical criminology), for example, situate thought, discussion, theorizing and policy in distinct sets of signifiers. Deleuze and Guattari also tell us that each regime of signs is replete with "order words," that is, master signifiers, as in Lacan, which convey unilateral information that insists. They tend to produce a specific effect. They produce what Deleuze and Guattari refer to as "incorporeal transformations."[17] Lacan goes so far as to say

---

16. Deleuze cites Foucault's *Archaeology of Knowledge*, *The Birth of the Clinic*, *Discipline and Punish*, and *The History of Sexuality* as inspirational for his development of stratification.

17. The phenomenon of "secondary deviance" posited by labeling theory is but one instance, a process by which a labeled person begins to become that very label.

that language, indeed, speaks the subject. Penal law, for example, allows particular discursive constructions of delinquency (Deleuze, 1988: 47). Lawyers who insert themselves in the language of law are dictated by the meanings that are already there; they do not create new meanings (appellate decisions do). As Whorf (1964) has reminded us, language carves up the cosmos into particular categories and the users of this language are predisposed to certain ways of seeing and creating reality.

Each historically formed stratum, then, is a composite, a distribution, of these two elements: visibilities and sayabilities, forms of content and forms of expression. In a quantum holographic reconceptualization, visibilities are possibilities extracted from the implicate order; sayabilities are accommodative to the explicate order and are possible expressions of visibilities. Each socius, then, can be characterized by its strata or constellations of visibilities and sayabilities. These two forms are mutually constitutive, or in "reciprocal presupposition"; one cannot be reduced to the other. In other words, they are relatively autonomous spheres, but yet mutually interpenetrate each other. As Deleuze (1998: 62–63) says, "prison substitutes another person for the penal delinquent and, as a result of this substitution, produces or reproduces delinquency, at the same time as the law produces and reproduces prisoners." Penal law, in other words, produces delinquency which by definition leads to prisoners; prison objectifies identities as delinquents continuing beyond incarceration (no decertification process) and hence produces again, the subject of penal law. They are in reciprocal presupposition. As Deleuze (ibid., 33) once again says, "penal law still leads back to prison and provides prisoners, while prison continues to reproduce delinquency, make it an 'object,' and realize the aims which penal law had conceived ..." Said in yet another way, the production of the social and psychic reality of delinquency has two mutually reinforcing sources in feedback loops. With Bradley, we see this process as producing both enduring frequency interference patterns or spectra and everyday momentarily constructed logons, quantum information that assures longevity and potential effects. Constant regulation maintains fidelity in communication. It produces harmonic resonance and continued belief and investment in our contemporary defined social reality of crime. It also produces resistance to innovative practices such as a possible transformative justice. Foucault (1977: 282; see also Deleuze, 1988: 33) has suggested a police-prison-delinquency circuit which produces our social reality of crime.

Deleuze's reconceptualization of Foucault's work further specifies the difference between "form" and "function." "Form" organizes matter—we reconceptualize as a result of collapsing a collective wave function producing instantiations, or the real; thus producing prisons, schools, army barracks, and hospitals.

And "form finalizes function and gives them aims" (Deleuze, 1988: 33), i.e. punishment, education, training, and care. "Function" is what produces a result of some sort. Function, for example, since the discovery of the prison, is panopticism (Foucault, 1977), a hierarchal and asymmetrical method of surveillance, separation, training, and normalization. These are the disciplinary mechanisms that are constituted by the term function. The prison-form thus can be conceived of as a pilot wave shaping a constellation of logons generated within a noosphere and its system of noosigns, producing incorporeal transformations. The repetition of the collective collapse of the wave function, the "social wave function" according to Wendt (2010: 296),[18] and adaptive phase conjugation (resonance) produce an ongoing social reality of law-crime-punishment-prisoners in a perpetual reinforcing feedback loop. Panopticism, as a function, has also taken hold not only in the prison-form, but in education, military, hospitals, and workplace as a dominant mechanism for incorporeal transformations. It is a society-wide pilot wave acting on multiplicities producing tendencies toward modal types. Subjects, following Wendt (2010) and Bradley (1998) are "tuned in" to the wave frequencies prevalent in the holofield that reflect this social reality construction. They remain quantumly entangled via "sayabilities," common forms of expression. Neophytes or uninitiated, enter this domain and are quickly disciplined into the appropriate ways of "seeing" and "saying" by phase conjugation dynamics. To follow and embellish Wendt (ibid.), active subjects ("monads") are those who through attention are "in tune" with the prison-form, allowing for the implicate order (potential "visibilities") to become explicate ("sayabilities"); whereas, passive subjects ("monads") not sharing more conscious awareness of the prison-form remain nevertheless quantumly entangled within the implicate order and can, via attention, become in tune (phase conjugation, resonance) and hence instantiate (explicate order) circumscribed logons embedded in the holofield thus seeing and acting within a common social reality of the penal law-prison assemblage.

3. Abstract Machines

Each assemblage can be mapped as a diagram, a statement about its internal virtual vector field of forces of possibilities (Deleuze, 1988: 36–44; Delanda, 2002: 30–44). A diagram maps: virtual vectors (latent forces that are at work); tendencies (potential directions measured probabilistically); trajecto-

18. Wendt (2010: 297) has said that the "social wave functions entangle many individuals quantum mechanically ..." Each person is "embedded in countless social wave functions," but remain relatively "passive" in most in particular moments (ibid., 298).

ries (directions of movement); singularities (moments of unpredictability); and attractors (the location to which dynamic systems tend). Said in another way, each assemblage is the source of virtual vectors. These "vectors of tendencies and becomings ... take form along the lines traced by this virtual function or abstract machine" (Bogue, 2003b: 83). A phase map traces moment by moment changes in the movements of elements defined on the map (see Delanda, 2002: 30–44). A phase portrait is a completed sequence of trajectories. It is an actualization or instantiation of a real from the virtual possible. Phase transitions take place at critical threshold values of intensity (ibid., 45). Each possibility space has degrees of freedom (potential actualizations, or "possible histories," Delanda, ibid., 35) and is mobilized at any moment in differing initial conditions. Ever slight perturbation in initial conditions can propel trajectories in divergent directions (histories). A diagram, then, maps what is sometimes referred to as possibilities space. An abstract machine, a particular function, is the logic of the diagram playing itself out with real effects. It produces "cuts" in the otherwise flowing energy field. It has also been identified as a quasi-causal operator; that is, it produces effects in the form of overdetermination theorized by those opposed to a strict Marxian interpretation of economic determinism (Deleuze and Guattari, 1987: 90; Patton, 2000: 45).

An abstract machine has a "piloting" function (Deleuze and Guattari, 1987: 142; Bogue, 2003b: 83–84; Patton, 2000: 45). It organizes energy flow into distinct configurations and effects, or occasions of experiences, to use Whitehead's terminology. "It steers the quantum flow ... and emits new quanta" (Deleuze and Guattari, 1987: 223). Thus, for example, we can conceive of the "machine-prison" (Deleuze, 1988: 39) guided in its everyday functioning by panopticism (Foucault, 1977). It produces visibilities and sayabilities, forms and substances. Its disciplinary mechanisms (observation, categorization, normalization) take raw bodies and effectuates an incorporeal transformation. In other words, configurations of bodily forces are realigned in a particular direction. Similarly, a "law machine" (Bogue, 2003b: 78–86) is what instantiates various forms of justice, be it by way of what Max Weber (1978) referred to as formal rationality, substantive rationality, substantive irrationality, or formal irrationality. Deleuze and Guattari's (1986) critical analysis of Kafka's *The Trial* suggests that the law machine is a power machine exercised at each point of its functioning, tending toward creating docile subjects. But embedded are virtual undercurrents that offer a potential dismantling and a re-assembling of these circuits of power (Deleuze and Guattari, 1987: 88–89; Bogue, 2003b: 78–86).

Two general forms of abstract machines exist. An abstract machine of "overcoding" tends toward the production of axioms from which syllogisms spring leading to constructions of truths. An abstract machine of mutation, however, deterritorializes molar forms producing "lines of flight." Virtual "lines of flight" emanating from the abstract machine and its axioms remain "tendencies, becomings, directions of movement toward potential actualizations of various sorts" (Bogue, 2003b: 83). They produce change, mutations, metamorphosis. Two forms of deterritorialization can take hold: "relative" in which only the instantiated real ("molar") is effected, or "absolute" in which change takes place at the virtual ("molecular") level. These lines of flight, however, can turn out to have good or bad consequences (e.g., increased or decreased capacities to effect and be affected) and hence must be ongoingly assessed for their value (Deleuze, 1995: 173; Patton, 2000: 8, 66–67, 87). Deleuze and Guattari (1987, 510) cite absolute deterritorialization and its lines of flight that can be the basis of the "new earth," a radically new socius where ever increasing capacities to effect and be affected are ubiquitous. Accordingly, their strategy for social change advocates a "war machine," not in the traditional form of understanding, but one that actively produces social change, which produces an absolute deterritorialization, lines of flight that lead to mutation and metamorphosis. We will have more on its relevance toward a transformative justice in Chapter 6.

Quantum holography offers to reinvigorate Deleuze and Guattari's radical reconceptualizations of the organization of a socius. Much of their work is implicitly from this ontological perspective. We can see how the ambient ZPF, a cosmos in vibratory states of existence, with flows and fluctuations of energy, can be "cut" up into relatively stable social forms, blocs of spacetime with degrees of stability. A diagram maps virtual forces and offers potentialities along more probabilistic pathways. This map scripts the frequency waves and frequency interference patterns, each of which conveys information. It is, in our reformulation, equivalent to Schrödinger's probability wave function. The abstract machine is what gives effect to these configurations, producing the real from the virtual. As a pilot wave, we are reminded of de Broglie (1924) who had argued for a material form of the wave function that was unfortunately, in its early publication, given scant notice. But with Bohm's (1983) work especially, the pilot wave begins to take on significance in the literature. Here, Bohm's pilot wave can be conceptualized as a line of flight embedded with quantum holographic information. It tends to generate the explicate order from the implicate order traced by the diagram. As an emerging explicate order, visibilities are generated, reflecting but one form of a collective collapse of the wave function. Sayabilities are linguistic and nonlinguistic expressions of this emerging field. With the development of regimes of signs discursive subject positions are offered in which an "I" can take residence in speaking. Foucault's (1994) "technologies of the self," an examination of historical forms of self-expression, provide the background ego, which provides further constraints to the emerging "I." Assemblages are the manifest "sub totalities" to which Bohm refers. Interactions within, between, and among assemblages produce an actual occasion, an event (Whitehead, 1925, 1929).

Quantum Holography, Ecological Niches, and Critical Criminology

In this section we want to incorporate our previous understanding in a model that incorporates quantum holography, Bergson's fundamental insights, Deleuze and Guattari's development of assemblages and strata, and Gibson's notion of ecological niches. We will include four applications of this integration in the final section of this chapter. In the next chapter we will return in more detail to the question of consciousness and agency. In our final two chapters we will offer several more detailed applications to criminology, law and social justice. Let us first integrate Gibson's work with Deleuze, Guattari, Bergson, and quantum holography.

1. Quantum Holography and Ecological Niches

Noted commentators (see especially Robbins, 2000, 2006, 2010, 2012) have suggested that Gibson's ecological psychology approach has affinities with quantum holography and Bergson. Gibson's key works include *The Ecological Approach to Visual Perception* (1979), *The Senses Considered as Perceptual Systems* (1966), and *Reasons for Realism* (1982). They lay out a theory of perception that is based on invariant structures (i.e., consistencies) that populate a field. These are "affordances," or particular possibilities offered by an environment, and are carried as information by ambient light in an ecological locale (1982: 73). His "ecological optics" approach has it that ambient light can be considered an "ambient optic array." "It is an *array* because the variation of intensities produces a differential arrangement within the total field of structured light. It is an *optic* array because of the differential arrangement, constituted by the variations in intensity, provides an active, mobile perceiver with information about the environment" (Braund, 2008: 133). In this model, a person perceives these affordances directly without any intermediary step. An "invariant" is something that remains consistent even in changing situations. An "affordance" is "what the environment offers the animal, what it provides or furnishes" (Gibson, 1966: 285; see also Robbins, 2006: 369–371;

2010: 514–515; 2012: 27–30, 71–115). These are, in short, structured possibilities or constrained opportunities. Perception of these affordances, for Gibson, is "direct." Of course, as Goldstein (1981: 193) rightfully states, Gibson gave no account to the learning process which associates an object with what it affords. An object, in other words, can have a different meaning to different entities; as we saw with Delanda's (2006) suggestion, each person has a differential capacity to effect and be affected depending on unique threshold values. An "ecological niche," for Gibson, is a constellation of affordances (Gibson, 1986: 128). In other words, it is an area within which a particular range of opportunities is found.

Gibson's model has received some criticism as a stand-alone, and some (Robbins, 2000, 2006) have made a compelling case that it only makes sense if we assume Bergson's quantum holographic suggestive model. In other words, if we accept the notion of the vibratory holofield, a domain of frequency waves constituting information that is accessible with the correct reference wave that reconstitutes information in visual 3-dimensional form, then Gibson's affordances as ambient light, that is "ambient optic array," as informational, is accommodative and more compelling.

2. Bergson and Affordances

Integrating Bergson with insights from Gibson, attentive recognition (Bergson, 2002) is where affordances offered by a particular environment are problematized whereby a zone of indetermination makes its appearance within which the "circuit diagram" is activated. Bergson's (2002) circuit diagram represents the search and activation of various memory-images that may provide a match with an engaged incoming image.[19] Incoming affordances, that is, vibratory

19. Bergson's (2002: 105) circuit diagram suggests that when confronted with an encounter, the percipient begins a rapid search for past memory images that might match the incoming image. In his words, "we maintain ... that reflective perception is a circuit, in which all the elements including the perceived object itself, hold each other in a state of mutual tension as in an electric circuit so that no disturbance starting from the object can stop on its way and remain in the depth of the mind: it must always find its way back to the object from where it proceeds." What a "state of mutual tension" suggests is juxtaposed images ("superposed"), each of which could be collapsed representing a recognition image that duplicates the incoming image. This, we shall see, has affinities with Penrose and Hameroff's (2011) idea of a momentary bubble that develops in the fabric of spacetime, which, with the appropriate level of intensity, collapses, eliminating the bubble formed, and instantiating an image that is now seen as reality.

waves with holographic information, are received by the percipient's transducers and interact with the percipient's signature wave. Between receiving an affordance and emitting a wave reflecting both the percipient's unique signature wave and the object-bearing wave, the percipient engages various possible virtual images (Bergson's circuit diagram) until a close approximation is found, which then becomes encoded in the outgoing wave. The standing wave (interference wave) as a composite of the signature wave of the percipient and the object-bearing wave is returned as an altered wave to the source which then appears where it is located, "out there." Consciousness, then, is *in* the object, not in the brain. The object activated then returns another wave back to the percipient to which his/her sensory motor apparatus responds with motoric action. It is the unique signature wave of a percipient being in tune (resonating) with affordances offered by a particular environment that produces adaptive resonance, or phase conjugation. The activated zone of indetermination is the space within which an affordance is matched to various memory-images; and, where images are matched, we have recollection-images. Inattentive recognition, on the other hand, is where resonance is not taking place at the more conscious level; it is "direct perception" and plays out the information embedded within the optic array. This integration can also be greatly benefited by referencing Wendt's (2010) distinction between "active" and "passive monads," where the former maps to attentive recognition, the latter to inattentive recognition.

3. Deleuze, Guattari and Affordances

To map Deleuze and Guattari's (1987) and Deleuze's (1988) work, invariant structures are assemblages, either molar or molecular, that often appear in layers as strata. These are virtual vector fields that provide affordances or quasi-pilot waves that traverse the field. They are information-bearing quantum holographic waves. These ambient lines of flight, as wave forms permeating the social field, are accessible by agents. This access is via phase conjugation, a resonance between the information bearing wave and the signature wave of the agent. Their interaction produces, in reciprocal fashion, appearances, the real from the virtual. Those "in tune" with the spectra that populate this field, that is their signature wave as a reference wave can access information in the holofield, can construct 3-dimensional imagery "out there" where they are located via phase conjugation, or resonance. Or, to approach it by referencing Wendt (2010), invariant structures are information-bearing "social wave functions" to which "active monads" can become in tune (resonate). By a common repetitive collapse of the wave function, we witness particular instantiations—what we "see" as the real and what is perceived as a common understanding.

The field of ecological psychology needs these further integrations for a bona fide statement on direct perceptions within ecological niches.

4. Ecological Niches as Bounded Spheres

In this spirit, ecological niches can be viewed as bounded spheres (noospheres) or a habitus (Bourdieu, 1992), or more generally, a possibility space on whose boundary surface all information is encoded. This information can be mapped as a diagram, whose real life effects are produced by an abstract machine. An abstract machine, recall, acts like a mathematical function, it produces particular results. Building on Deleuze and Guattari, an abstract machine is the source of affordances in a particular niche. Affordances can also be likened to Bourdieu's (1992) "dispositions," or tendencies. Each bounded sphere can be conceptualized as populated by unique ambient affordances originating from the boundary surface area enclosing a volume. It is basically a quantum wave function, $|A_i>$. This is not to say that emerging diagrams and abstract machines do not also populate the spacetime bulk, the interior of bounded areas; in fact, we would speculate, mutation and metamorphosis has more to do with the internal dialectic between forces of territorialization and deterritorialization within the bulk and the outcome registered on the boundary surface. The "refrain" (Deleuze and Guattari, 1987), reflecting order out of disorder, the becoming-of-order, can also be the source of new harmonic resonances that take on more stable form, be it in the more rigid molar, or dissipative molecular assemblage form. We would also speculate that there must be a process by which these re-appearing patterns of energy are further encoded on the boundary surface in one less dimension and become the basis of subsequent affordances; a theory is in need of development as to the interaction between this double inscription. The precise physical dynamics of this reinscription remains unclear at this point.[20] In inattentive recognition (which connects with "direct perception") we would hypothesize that dominant "templates of action" or sensorimotor schemas[21] will be more likely instantiated. In attentive recognition, however, direct perception and routine action are not

---

20. We are reminded of Freud's work in *Interpretation of Dreams* and his analysis of a double inscription of sense data, once in the mnemic system, and then again in the unconscious system.

21. Templates for action, or sensorimotor schemas can be envisioned as patterned motor action. Consider much of our everyday activity that seems more "on remote," from turning a door knob, to changing gears in a manual drive car, to pouring a cup of coffee, and so on.

linear; rather, they are momentary potential occasions for mutation and metamorphoses. In other words, the collapse of the wave function could produce novel instantiations.

Quantum Holography at Work

Let's take four examples of quantum holography at work. We will apply our thoughts to: capital logic and commodity/legal fetishism; Foucault's panopticism; classical subcultural theory in criminology; and to terrorist organizations. In our final two chapters we will return to applications in critical criminology as well as to law and social justice.

1. Capture and Subjectification

The notion of capital logic is often brought up in Marxist literature. This logic was developed by Marx (1992) in his notion of commodity fetishism, specifically, drawing from the first hundred pages of *Capital* (1967). Pashukanis (2002; Milovanovic, 2002), building on Marx, was to show there was a homology, a similar development, between commodity fetishism and legal fetishism. Just as the use-value[22] of an object is, via constant transfer of commodities at the competitive capitalist market place, transformed, "behind people's backs," into exchange-values, or abstractions, we also witness the emergence of particular phenomenal forms: the appearance of free will, proprietary interests, and equality. Together, the abstraction is formed of the legal (juridic) subject with inherent rights. This hidden process is given "idealized expression" by lawyers. Thereafter, these idealized expressions become constituent elements of noosigns within a capital logic driven socius (noosphere). Deleuze and Guattari would say that commodity exchange produces visibilities (i.e., use values being converted into exchange value), and sayabilities (i.e., idealized expressions).

Following Deleuze and Guattari (1987: 223), political economy is a system of "capture," that is, a system that specifies the appropriate crystallization of bodily forces, in the form of absolutes, or axioms. This axiomatic system (diagram), or territorialized system, provides a linear process for creating truth

22. Use values have two components: the amount of labor used in its production, and the needs to which it provides satisfaction. Both represent differences, or qualitative relationships. Exchange value represents quantitative relationships, ratios of exchange, equivalent relations.

by way of syllogistic reasoning and deductive logic. In our reconceptualization, this axiomatic system is codified on the boundary surface and is the source of affordances, or an ambient optic array, which illuminates the interior volume. It is also a quantum wave function that tends, due to the abstract machine, to collapse in system-directed forms. Capitalist logic, as a mapped diagram, is a societal wide abstract machine that continuously creates lines of flight or pilot waves, much like solitons, within the socius. It is what Gibson called an "invariant structure" (see also Robbins, 2006: 369–371; 2010: 514–515; 2012: 27–30, 71–115); that is, it remains constant even with changing activities in the socius. Structural invariances are, reconceptualized, common frequency interference patterns (spectra). They can also be conceived as a "social wave function" (Wendt, 2010: 297) that tends to entangle agents quantum mechanically. Agents within the volume have access to these affordances, these ambient optic arrays that are information bearing by way of phase conjugation. Three dimensional constructions follow resonance dynamics. Agents, in short, receive affordances which are responded to in more passive form (Bergsons' inattentive recognition) generating direct perception and resonance; or, in a more active way (Bergson's attentive recognition), generating a search for other actualizations. The former tend to be the modal forms under the influence of an abstract machine of capture that collapses the wave function of possibilities to that of the logic of molar assemblages. The ambient holofield is populated by assemblages, or constellations of frequency patterns that are available for actualization by way of a reference wave (Bradley, 1998). The notion of a juridic subject, a rational subject in law, and notions of free will and formal equality are reflected by a quantum lexical structure (noosigns) that provides signifiers and especially master signifiers or "order words" that are in reciprocal presupposition with capital logic. Corporeal transformations are but tendencies or "dispositions" (Bourdieu, 1992) of these virtual vectors existing in the ZPF.

The effects are brought out by Deleuze and Guattari's example of identity constructions. They show, for example (1987: 137–139), how beginning with some master signifiers, order words, or a "point of subjectification"—an axiomatic system—produce a "subject of enunciation," a temporary crystallization of bodily forces. This leads to an emergent "subject of the statement," an "I" that takes up residence in appropriate discursive subject positions in constructing narratives. And it is the "I," the subject of the statement, "into which the subject of enunciation recoils" (ibid., 137). Said in a quantum holographic way, the social wave function emanating from an axiomatic system which is an abstract machine producing distinct optic arrays, or pilot waves, will tend to collapse and be given a repetitive instantiation—the perception of free will and formal equality, an idealized rational legal subject in law ("reasonable man in law"). Particular instantiations have nonlocal effects. It generalizes to the social field. It is harmonic resonance (phase conjugation) with frequency interference patterns that have attained relative stability because of the continuous reconstructions by inattentive recognition (e.g., Wendt's passive monads). Quantum information in the form of spectra is continuously given further fidelity in everyday reconstructive use. In other words, consistent with constitutive theory, suggested is that order is not something objective, "out there," but is reconstructed by agents in their everyday inattentive recognition practices. Order is maintained to the degree that the disciplinary structure mapped as a diagram on the boundary surface of a socius and played out by a correlative abstract machine continuously produces an ambient optic array to which agents access via phase conjugation and resonance.

In another context, we can conceptualize how Deleuze's Nietzschean (1989) inspired idea of the "man of truth," an agent who insists on being in touch (resonating) with universal truths, and the idea that "truth" has some objective basis that can be discovered or made visible, are both realizations of affordances producing incorporeal transformations. This process is uneven, however, partly due to the ubiquitous Heisenberg uncertainty principle. Degrees of uncertainty remain inescapable. We shall also see in our final chapter, that transformative justice can make use of Deleuze and Guattari's notion of an alternative machine, the "war machine," an "affirmative nomadology," that effectuates deterritorialization and lignes of flight producing a "new earth," an alternative social (dis)order. That is, new logons reflecting molecular assemblages can become ubiquitous which remain always in dynamic states, resisting closure and stases.

2. Panopticism and Disciplinary Mechanisms

A second example can be presented by reformulating Foucault's (1977) notion of panopticism and disciplinary mechanisms. Panopticism,[23] as a function, can be mapped as a diagram that is encoded on the boundary surface of a socius. It is an axiomatic system embedded on this surface. An abstract machine is what gives particular form to the function creating bodies (incorporeal transformations) in the social field. An optic array begins here and is disseminated throughout the enclosed volume, the spacetime world we experience. This optic array embeds information quantum holographically. It is characterized by particular frequencies waves that subject-agents can ac-

23. Foucault's (1977) classic study of the evolution of prisons shows how panopticism is a system of asymmetrical surveillance and control. Disciplinary mechanisms normalize behavior, that is subject it to some preconceived notion of acceptable order.

cess via phase conjugation. The world we perceive is but these holographic projections with which all are in tune (resonate). In Bradley's language, the interaction between flux and control (disciplinary mechanisms) produce spectra populating the ambient holofield that await activation (instantiated as logons). Social policy makers, mainstream scholars, prisoners, teachers, agents of social control, and other subject-agents uniformly collapse this social wave function producing collective and modal forms of instantiations. This uniformity represents but one possible instantiation of the real from the virtual possibilities. Again, this process is uneven due to Heisenberg's uncertainty principle behind the wherewithal of everyday construction of logons as Bradley (1998, 2011, 2011) perceptively has demonstrated. However, tendencies to produce molar assemblages and strata remain logical deductions of this axiomatic system. Consider, for example, the liberal response to the failures of our criminal justice system, the restorative justice approach. Subject to a critical examination, it is suspect as to its embedded abstract machine that tends toward the reproduction of compliant subjects. Woolford (2009: 85) has well stated the process in the production of compliant, passive subjects, "through the cues and interventions they use to remind participants of how they should be communicating. Through a great deal of preparation and frequent reminders, participants become disciplined to norms of effective listening and interest-based negotiations." In Foucault's (1994) late work on the "technologies of the self," focusing on how particular historical forms of constructing appropriate selves are stabilized, we see, Woolford tells us, how restorative justice is a form of "soft coercion" since subjects are taught to be ultimately responsible for their actions, and to see this as unquestionable (Woolford, 2009: 86–87; Pavlich, 1995). However, inherent tendencies of deterritorialization are ubiquitous, the very nature of assemblages. It remains an activist's task to engender deterritorialization and a reterritorialization that tends to the production of molecular assemblages which reflect greater potentials for expression (an alternative wave function), and for increasing capacities to effect and be affected. Restorative justice just won't do; we need a transformative justice which we will more fully develop in Chapter 6.

3. Subcultural Theory in Criminology

Our third example of application of quantum holography can be applied to more traditional subcultural theory in criminology. Let us briefly address a few. These are classics in the field and resurface (accessed in the ZPF!) continuously in contemporary theorizing. Therefore, it is a productive exercise to go to the source of much current conventional thought in criminology.

Walter Miller's (1958) classic theory, "Lower Class Culture as a Generating Milieu of Gang Delinquency," presents several "focal concerns" that are part of the subcultural life of "lower" classes. Notwithstanding the pejorative notion of "lower" class and its many critiques, Miller is indicating how invariant structures have ongoing influence on those who frequent this spacetime. In other words, this "generating milieu" is an ambient holofield offering spectra that codify "focal concerns." Fidelity is maintained in their continuous activation and use. Said in another way, within this boundary region we can identify characteristic affordances (focal concerns) and dispositions (tendencies, "templates for action"). In the everyday commonsense understandings of reality and actions that follow, inattentive recognition and direct perceptions are ongoing modal forms of action. Direct perceptions (Gibson, 1979) are more direct harmonic resonances producing everyday 3-dimensional perception-images. The wave function representing the many possibilities of instantiation are collapsed in a repetitive way resulting in patterned behavior (periodic and cyclic attractors, i.e., focal concerns). Miller has provided idealized expression of this dynamic without providing the basis of its physicality.

We also noted earlier that gang workers, such as those employed by programs employing former gang members as intervention agents ("interrupters"), called "Cure Violence," attempt to deterritorialize molar assemblages in this milieu by encouraging youth at risk to construct a different set of logons (explanations). By these intervention practices they collapse the wave function into forms that lead to less violence.

Cloward and Ohlin's (1966) classic study of three subcultures of crime as a response to blocked opportunities and strain, can also be reconceptualized in terms of quantum holography in showing that these subcultures create bounded areas within which certain forms of expressive or instrumental behavior predominate, which, in turn are based on the logons constructed and how the optic array is collapsed producing the real. Of course, notwithstanding is criticism levied against their model in that two or even three of these forms can be found even in the same "specialized environment." But this also brings out the idea of relatively autonomous nested hierarchies of bounded spaces. That is, optic arrays may cross bounded spheres (i.e., "quantum tunneling") having distinct nonlocal effects. Walter Miller, for example, has noted how the "lower class subculture" has a substantial influence on the American public.

At a more micro level, consider Cohen's classic study of *Delinquent Boys* (1955) which shows that delinquency often begins in group or gang settings. As he informs us, "deviant as well as non-deviant behavior is typically not contrived within the solitary individual psyche, but is a part of a collaborative social activity, in which the things that other people say and do give meaning, value and effect to one's own behavior" (ibid., 8). Again, these settings can be considered as shared bounded regions on which coded information (diagram) about ongoing activity is stored. An abstract machine is the playing out of this logic in the form of an ambient optic array, or pilot wave that is a quantum holographically bearing information wave that is accessible by subject-agents within the enclosed spacetime volume (milieu). The "collaborative social activity" suggests that harmonic resonance is ongoing, producing high fidelity, and a characteristic form of instantiation.

Another example would be with the work of the Chicago School, particularly the ecological approach by Shaw and McKay (1969). Their classic study divided the city in terms of concentric circles indicating criminogenic "transitional zones." Reconceptualized, these are bounded regions with distinct diagrams (affordances), abstract machines (dispositions), optic arrays (quantum holographically bearing information waves), pilot waves (lines of flight), and ubiquitously replete with ongoing constructions of patterned spectra—all accessible (nonlocality) in their individual, momentarily constructions of logons, by those in tune within this milieu/niche.

We hasten to add, we are not attempting to support the above theories. We merely want to show that they can be reconceptualized along quantum holographic principles and thus are provided a physicality for their concepts and dynamics. Our analysis is in the spirit of "naturalism" (Matza, 1969)—understanding, not correctionalism. Also, the implicit determinism of these approaches can be overcome by appreciating the inherent indeterminacy in quantum holography. At best, we have a probability function, a probability wave that is a virtual "cloud of possibilities" from which the real is instantiated. Future research on these classic theorists, then, could entertain a quantum holographic reconceptualization in order to have a better appreciation of the nature of the dynamic phenomena at play. Clearly, other classic theories could point toward alternative constructions, such as we find in the work of Matza's *Delinquency and Drift* (1960) and in his *Becoming Deviant* (1969). These two works question a deterministic model and hence bring out Heisenberg's uncertainty principle, as well as Deleuze and Guattari's notion of deterritorialization.

For Matza (1960), a critical variable is the "situation of company," a group of juveniles, say on some street corner. At these occasions a shared bounded space exists, which we see as a relatively autonomous sphere nested within many others. He also posits various "techniques of neutralization" available to the juvenile in this niche. This is a quantum lexical structure encoded on this boundary surface and reflected in activity within the volume (noosphere) in everyday constructions of logons. Neutralizations are core noosigns accessible for in-tuned juveniles. To borrow from Wendt (2010), within the noosphere, a social wave function exists with which active agents are in tune; that is, agents ongoingly establish harmonic resonance (informational channels) from which the repetitive collapse of the wave function produces particular patterned instantiations (delinquent subcultures). That is to say, the operative diagram functioning through its abstract machine generates pilot waves (or affordances) that energize the holofield, the bounded space of a street corner setting. The juvenile's unique and ongoingly emitted signature wave interacts with this pilot wave in constructing logons within the holofield. In the "situation of company" what is produced is "a self-reinforcing collective field of coherent bio-emotional energy" (Bradley (2007: 87); that is, each juvenile finds him/herself in "harmony with the resonant frequency of the group as a whole" (ibid.). This heightened resonance produces both solidarity as well as more intensified harmonic resonance; in other words, an information channel is opened which offers neutralizations, signifiers, and plans of action (dispositions, templates for action). The "trigger," following Delanda's notion of catalytic causation (see Chapter 2), is a felt state of "being pushed around," which has differential thresholds and effects with different persons.

In a passive mode (Wendt's "passive monad," 2010: 300), or inattentive recognition (Bergson, 2002), we once again hypothesize that direct perceptions (Gibson, 1979) and the playing out of recurring dispositions (Bourdieu, 1992) is a more likely event. A common quantum lexical structure provides a basis for an insertion of an "I" in discursive constructions from which follow possible circumscribed loci of perception-images and greatly circumscribed action patterns (point and cyclic attractors). But logons, because of their overlapping nature, provide a degree of uncertainty; hence Matza's novel idea that juveniles "oscillate" and drift in and out of delinquency. It is not a deterministic process. Labeling theory, of course, would suggest that agents of social control intervening in the juvenile's momentary and episodic delinquency will tend to stabilize a uniform collapse of his/her wave function; in short, identities are solidified, and secondary deviance has a heightened probability of following.

4. Terrorist Cells, Boundaries and Phase Conjugation

Our final example concerns terrorism. An explicit model of the crime of terrorism has been applied by Bradley (2010, 2011) that we previously explicated. Let us now extend his approach by incorporating the quantum holography principle developed from studies of the black hole and how information is stored on boundary surfaces. Bradley's model suggests that the notion of a bounded area can also be applied not only to secret and non-secret groups, but also to networks established via the internet. Bradley's unique contribution is in linking two holographic approaches and demonstrating how harmonic resonance takes place between agents and the unique signature wave of the secret society. This produces solidarity, reinforces group boundaries, and provides high fidelity in logons constructed. He uses classical holographic principles (Gabor, 1948) to show how recurring activity of members is imaged as a hologram and is encoded as Fourier descriptors in each participant's brains. These we can map to Gibson's notion of structural invariances and affordances, as well as Deleuze and Guattaris' strata formations. They are the basis of stability. He then uses quantum holographic principles (Gabor, 1946) to show how subsequently this configured activity is accessed by members in their everyday activity. In other words, with each interaction information is created which is quantum holographic in form (i.e., spacetime constrained holograms or quanta of information, Gabor, 1946). These logons are then matched with the group's unique and more stable (configural) signature wave. The resonance, or phase conjugation, is experienced as a felt commonality, or solidarity. This produces constancy and identity out of the otherwise flux of energy flow.

Bradley relies on Pribram's notion that spectral group identity waves are incorporated in the brain as Fourier descriptors, whereas, reformulating, we follow a double-inscription holographic model, drawing from: (1) Mitchell and Laszlo in suggesting it is encoded in the holofield, the ZPF; as well as from (2) the quantum holographic model developed by 't Hooft, Susskind, Maldacena, Bekenstein and Bousso suggesting boundary surfaces encode all information on happenings in the volume, or bulk. The boundary surface area can be envisioned as a form of photographic storage plates we explained in the double-slit experiments (see Chapter 2). The boundary surface is the source of diagrams (affordances), abstract machines (functions, dispositions), ambient light (encoded with quantum information), and pilot waves (acting much like solitons); the spacetime bulk is a repository of more dynamic, everyday activity (logons) that creates and reconstitutes spectra anew. Bradley makes the point that counter to attempts to locate terrorist cells by way of intercepting messages a better way would be to identify the signature pattern of terrorist cells and their members. Be that as it may, however, still needed is a statement on agency (to which we will return in the following chapter). Wendt, of course, does make a distinction between a more passive and a more active subject. He shows how agents can be in tune (resonance) or out of tune with the social wave function. Recall, too, Bergson makes a distinction between inattentive recognition and attentive recognition. In the passive form of inattentive recognition, affordances lead to "direct perception" (Gibson, 1979). In more active attentive recognition form, affordances and structural invariance's are subject to delay in response. We have seen, for example, that at this moment, for Bergson, a "zone of indeterminacy" appears at which time the "circuit diagram" is invoked whereby alternative perceptions and actions may follow. In the next chapter we will have more to say about the nature of agency as it relates to quantum holography.

Conclusion

We have covered some of the more salient concepts emerging from quantum holography. We saw that at least four significant variants of a quantum holographic approach have emerged. Each has underscored the primacy of information—its dynamic construction, storage, transmission, and interpretation. Phase conjugation was shown to be the key mechanism by which this process takes place. This chapter concerned a reconceptualization of much of Deleuze and Guattari's work along the lines of quantum holography and offers a new direction for critical criminology. We noted that much in their work implicitly, sometimes even quite explicitly is about quantum theory. Holographic theory is less implicit. Bergson's work has been more implicitly quantum holographic as Robbins (2000, 2006) has suggested. We have also seen that the work of Gibson on affordances can be incorporated. His notion of the agent as mostly engaged in direct perception was modified to show that this form of perception is accommodative to Bergson's notion of "inattentive recognition," not "attentive recognition," the latter providing the possibilities of innovation.

We concluded our chapter with some short applications of the concepts covered in this chapter. These are not definitive, but more in the form of snippets and first approximations that suggest the necessity of much more substantive analysis. We are now ready to delve deeper into the nature of agency itself.

Chapter 4

The Question of Agency: Quantum Holography and Schema QD

Introduction

Much of contemporary traditional and critical criminology not only finds itself grounded in the logic of Newtonian physics, but also lacks a bona fide statement on agency. A similar state is found in law and social justice research. By agency we mean the human being's abilities to develop innovative perception-images as a basis of consciousness and subsequent actions. Consciousness is generally viewed as awareness and its relation to self-awareness. The notion of a "will" and self-perceived feelings of a free will are components. We also follow Spinoza's idea of a subject's capacity to effect and being affected. These are essential dimensions of what we refer to as agency. David Chalmer's (1997, 1999)[1] classic analysis posed this as the "hard problem" of consciousness, that is, the question of the connection of the biological brain to subjective experiences. We resist reductionist approaches that relate agency to merely neural correlates of consciousness. For the most part, physicists acknowledge that consciousness studies are not within their domain. There has been a growing exceptions list. Von Newmann's (1932) classic and much-cited work on providing the mathematics for quantum mechanics, *The Mathematical Foundations of Quantum Mechanics*, is referred to as the Copenhagen interpretation. It is consciousness, he argues, that collapses the wave function producing the reality we observe. The wave function itself merely provides possibilities from which one is actualized. However, those to follow, at best, rely on a notion of some form of an "abstract ego" that is behind the observations that produces the collapse of the wave function.

1. See also: http://consc.net/papers/facing.html.

We relate the importance of this to how the "social reality of crime" is constructed (Quinney, 1970). Images created are the basis of identification and action, and thus we want to delve deeper into the quantum holographic nature of consciousness and what light it reflects on the construction of the social reality of crime. In this chapter we would like to show the relevance of quantum holography to the development of agency. We will review some relevant literature that begins this analysis, and build our own model that is a de-oedipalized quantum holographic adaptation of Jacques Lacan's (1977) inter- and intra-subjectively constituted schema R that we will refer to as schema QD.

The recent rush toward rational choice theory and its variants that are often tied to quantitative/empirical studies in criminology relegate agency to an insignificant dimension of efficacy. Positivism, too, fairs no better, with its relegation of the subject to forces that are said to propel him/her into action beyond his/her control. Law, too, finds itself confined to a restrictive notion of agency, the abstract juridic subject or normalized "reasonable man in law," with an occasional nod to internal or external forces such as in the case of the insanity defense, duress, entrapment defense, "station interrogations," and such formulations as the "product rule" and Miranda warning. Notwithstanding, of course, is how excusability is correlated with hierarchical location, whereby the scope of allowable conditions for justice rendering widen as we go upward in the class structure, and hence, more excusability (Milovanovic, 2003: 128; Kelman, 1981).

In social justice research all too often we are provided with an un-thought idealistic account of the inherent free nature of agency and desire to return to Nirvana. Some, in postmodernist research, would also render agency insignificant. Jean Paul Sartre's (1972: 37–57) critique of Bergson's anti-transcendental subject, a critique equally applicable to Deleuze and Guattari's work, does demand a better grasp of agency itself.[2] Our work here to follow not only advocates a subject-in-process, heading Sartre's well developed critique, but also suggests a diagram, a first approximation, of its major constitutive elements. Agency cannot be either conceptualized as transcendental nor as non-transcendental, but, in our view, somewhere in between. It is to quantum holographic theory we look in order to provide a better understanding of the complexities of agency.

Arguably, David Matza's work (1960, 1969) has been, and still remains, the most productive in integrating a notion of the subject in his explanation of

2. Sartre (1972: 40) asks, "How do virtually represented images suddenly manage to encompass an 'I' by becoming 'present'"?

crime and delinquency. His approach, particularly in his book, *Becoming Deviant*, suggests an active subject, which on occasion can be pacified; in this state of objectification, the subject may indeed simply play out the societal scripts laid out before him/her. Steve Lyng's work on edgework (1990, 2005) has turned homeostatic oriented research on its head, indicating that the subject is not prone to just adjusting to given conditions trying to minimize bodily tensions, but just the reverse, in order to activate higher states of agency. Moreover, this playing out of adrenaline highs is seductive. Jack Katz's *How Emotions Work* (1999) and Norman Denzin's *On Understanding Emotion* (2007) have also made considerable advances in indicating that emotionality, or affect, is a critical ingredient in agency, and that it is constituted intersubjectively. Thus for Lyng, Denzin and Katz, criminology is overly preoccupied with background factors (traditional examination of structural factors—e.g., unemployment, slum conditions, violent gangs, and various inequalities such as class, gender, race, etc.) and not enough with foreground factors (emotionality and affect). For each, the notion of "I think therefore I am" completely misses this critical dimension of subjectivity/agency (Katz, 344). More recent work by Ferrell *et al.* (2008) and Jock Young (2011) have expanded analysis of cultural criminology to include these components, and thereby are beginning to lay out a framework that better integrates agency into structural analysis. Jock Young's recent *Criminological Imagination (2011)*, following C.W.Mills' (1959) call for thinking imaginatively, lends itself to further integration or reformulation by the application of the conceptual tools of quantum holography. As it stands right now, perhaps cultural criminology offers one of the best hopes in contemporary criminological thought. But yet it still resists shifting away from underlying ontological assumptions rooted in Newtonian physics.

Arguably, "queer theory" with its emphasis on the idea that identities are not fixed, is on the borders of making a paradigm shift, even as it currently is exhausting the possible limits of theorizing given its mostly Newtonian ontological assumptions. Similarly, for LGBTQ theorizing (lesbian, gay, bisexual, transgendered, and questioning/queer). Perhaps the "Q" is most suggestive for alternative formulations of identities that have not yet appeared in the literature. "Questioning" is consistent with Deleuze's (1989) idea of a "people yet to come."

Finally, in our short comment on the contemporary criminology's understanding of agency, critical geography, and perhaps developments in a critical geographic criminology, or alternatively, a critical criminological geography identifies urban locations as bounded regions which find various correlates with disenfranchisements, inequalities, bias, and injustice. It lends itself to an integration of a viable notion of agency. Quantum holography is offered here as that essential ingredient necessary for reformulation and resurrection from Newtonian ontological assumptions.

Jacques Lacan, a revisionist Freudian, who's theorizing is overly based on "lack" as the basis of desire, with revisions, offers opportunities for clarifying the nature of a comprehensive inter- and intra-subjectively constituted agent. Even with Deleuze and Guattari's frontal assault on the oedipalized nature in their book *Anti-Oedipus*, they still acknowledge that many facets of Lacan's version of psychoanalysis have importance in understanding the subject. A deoedipalized subject not based on lack is essential for a new ontology. For example, it can be alternatively based on production (Deleuze and Guattari, 1987); on a "will to power" (not to be confused with a drive toward domination, but a drive toward greater empowerment), becoming and "affirmation" (Nietzsche, 2003; see also Deleuze, 1983); on Spinoza's (1991) joyful passions; or on an "élan vital" and ever evolving "creative evolution" (Bergson, 1998, 1999). Accordingly, we will modify Lacan's schema R, one of his central diagrams of the subject and its perception creating abilities, by incorporating alternative quantum holographic principles.

Let us first review groundbreaking developments on quantum consciousness rooted primarily in physics and bioneurophysics. We shall then move to an alternative construction of the subject/agent in the form of schema QD that is informed by both quantum and holography theory. We conclude with a suggestive direction in the relationship between political economy, a system of capture, and subjectification.

Quantum Consciousness

Two prominent approaches on a quantum consciousness have been articulated in physics and bioneurophysics. The first we will briefly summarize is by Roger Penrose and Stuart Hameroff (Hameroff, 2011, 2012; Hameroff and Penrose, 1998, 1996a, 1996b, 2007a, 2007b, 2010),[3] the second by Henry Stapp (2007, 2008, 2011, 2012, 2013a, 2013b).[4]

---

3. See Hameroff's extensive work in this area on his homepage: http://www.quantum-consciousness.org/.

4. Stapp's work in this area are listed on his homepage: http://www-physics.lbl.gov/~stapp/stappfiles.html. See also an overview of his work: http://www.informationphilosopher.com/solutions/scientists/stapp/.

1. Roger Penrose and Stuart Hameroff: Objective Orchestrated Reduction ("OR")

Roger Penrose the noted, influential physicist, and Stuart Hameroff, an anesthesiologist, have co-authored several articles (see the recent Penrose and Hameroff, 2012) that relate biological functioning to quantum states and their collapse. In their view, there occurs an "objective reduction" (OR) in the collapse of the wave function. Microtubules, the protein skeletal structure within neurons, are said to provide the basis of two or more image states. This superposition of states is quantum coherence. They apply an approach in physics called quantum gravity. At the moment of trying to make sense of external sense data, ambiguities generate a momentary separation, blister, or bulge in the very space-time fabric. Technically speaking, it produces a space-time curvature at the Planck scale, the smallest conceptualized space which is 10^{-33}cm. This can be conceptualized as two or more side by side planes representing possible states. This state of quantum coherence is inherently unstable and at a certain threshold value will objectively collapse ("decoherence") producing one image we perceive as the response to a particular incoming sense data. This is equivalent to saying the collapse of the wave function. This is what Whitehead would call "occasions of experience," or much like a sequence of film shots, an ongoing stream of consciousness. Said in another way, each occasion of experience is a quantum state reduction, or collapse of the wave function. At critical threshold values of intensity it is quantum gravity that objectively produces the collapse. What influences this collapse is nonlocal "hidden variables" that are said to guide the process. Moreover, this is a "noncomputable" process, defying any precision as to which direction the collapse will go. Instantiations, in other words, guided by qualia are never predetermined. They are neither random occurrences, nor algorithmically chosen. Each collapse produces a momentary state of conscious awareness, qualia or sensations and raw feelings. At the Planck scale it is theorized that space-time geometry is organized in terms of "spin networks," a spider-looking network of spin states.[5] Both philosophers Spinoza (1995) and Leibniz (1991) are inspirational in this theory as they provide for even elementary consciousness to all entities, or "monads" according to Leibniz. Whitehead is in accord, and his view could be conceptualized as suggesting a continuous flow of these events, or better, eventing. Thus for Hameroff (1998: 13) siding with this panpsychist approach, it is the "Planck scale spin networks

5. For some good visuals on spin networks, see: http://www.einstein-online.info/spotlights/spin_networks.

[that] encode proto-consciousness ('fundamental') experience (qualia) as well as Platonic values. Particular configurations of quantum spin geometry convey particular varieties of proto-conscious experience, meaning and aesthetics." Each objective reduction is equivalent to a conscious event; many, a stream of consciousness.

Let us take an example of this process, modified from Hameroff (1998: 14). You briefly are exposed to external sense data, someone's face. But you are not sure whether it is Fred, John, or Walter. Momentarily three states are superpositioned, are rendered quantumly coherent. For some 25 milliseconds there exists pre-conscious processing (ibid.), that is, microtubule are activated in particular configurations; for visual presentational purposes conceived as various figures on two-dimensional planes, side by side. At a critical threshold value of intensity, an objective reduction will occur collapsing the three possible states into one.[6] During this preconscious state, in other words, various possibilities are superpositioned. At a threshold, reduction, or collapse of the wave function produces a momentary state of consciousness, a particular classical state. Choice situations follow a similar pattern (ibid., 15). During the pre-conscious phase various possibilities find themselves in momentary quantum coherence. Quantum coherence is equivalent to quantum computation taking place. That is, the microtubule are being configured in various states. At some critical threshold value, objective reduction takes place. A particular choice is made (ibid.). In this view, the notion of "free will" actions is supported. "Quantum state reductions," Hameroff (2012: 2) states, "seem to involve temporal non-locality, able to refer quantum information both forward and backward in what we perceive as time, enabling real-time conscious causal action … OR can thus rescue free will."[7]

Backward time referral (see also the classic work by Libet, *et al.*, 1979, 1983) is also an important ingredient in the evolution of the final image (Libet, 1981,

---

6. Their formula is given as $E=h/t$: where E="gravitational self-energy" due to amount of tubulins superpositioned; h=Planck's constant over 2π; and t=time before superpositioned images collapse to one producing a "conscious moment." There is an inverse relation between E and t; small E (say an electron) reduces over millions of years, large E (say a cat) reduces within 10^{-37} seconds (Hameroff, n.d., 5–6; 1998: 2–3, 11; 2007a: 1037–1038; Hameroff and Penrose, 1996b; Penrose and Hameroff, 2011). Buddhist monks have been shown to produce over six million "moments" (conscious events) in 24 hours (Rospatt, 1995; see also Lutz et al., 2005).

7. Hameroff (1998: 15) has also said, "our 'free will' actions could be the net result of deterministic processes acted on by hidden quantum logic at each Orch OR [orchestrated objective reduction] event. This can explain why we generally do things in an orderly determinist fashion, but occasionally our actions or thoughts are surprising, even to ourselves." See also Penrose and Hameroff, 2011: 17.

1985, 1992, 2002, 2004, 2006; Bem et al., 2011; Bem, 2011; Zubairy, 2005; Eccles, 1985).[8] Sometimes it has been called "backward causation," "subjective backward referral," and "backward referral in time." Information is sent backward in time to add elements to the emerging picture. Repeated studies show that we begin to act before we are fully aware of our choices to act. Consider baseball or tennis players striking a ball and how much time they have to describe what the ball is doing (fast, slow, rising, curve, etc.) However, unlike Dennett (1991), Wegner (2002) and Huxley (1986) who conclude that conscious choice is thereby an illusion—"we are merely helpless spectators"—with the notion of backward time referral, consciousness is resituated as a central active player (Hameroff, 2006, 2007a, 2007b). Hameroff suggests hidden variables and qualia (felt states) and Penrose demonstrates the "noncomputability" in the operation of the brain which together offset any possibility of a strict, linear determinism. Each "objective reduction" refers quantum information to a previous time and thereby produces a more integrated perception-image (gestalt, réalité) early in the developmental process (Hameroff, 2006; Hameroff and Penrose, 2007a).

2. Henry Stapp: Abstract Ego and the Collapse of the Wave Function

A second key approach in the quantum consciousness tradition in physics has been provided by Henry Stapp (2007).[9] He distinguishes the mental (consciousness) from the physical (brain) realm. Consciousness can influence brain superposition states which then can be collapsed to one instantiation, one possibility amongst the co-existing possibilities.

Reality, for Stapp, appears as a series of collapses of the wave function. The observer, following Heisenberg, collapses the wave function by his/her observation or measurement. He follows Newmann's (1932) pragmatic approach, what later was to be called the Copenhagen school in quantum theorizing. For Newmann, as is the case with Stapp, the universe can be represented by a wave function, but is further composed of the totality of all the subjective collapses that each of us can effect by observation. His model of consciousness recognizes three processes that lead to the eventual instantiation, what we perceive as reality.

---

8. Radin (2006, 2013) has reviewed many empirical studies supporting this notion. His most recent work (2013) suggests connections to the literature in extrasensory perceptions and Eastern philosophy. For further empirical support and discussion of retrocausation, see the anthology by Sheehan (2006).

9. See his website for useful articles: http://www-physics.lbl.gov/~stapp/stappfiles.html.

Process 1 stands for the ability of an "abstract ego" to pose questions about its existence.[10] We ask nature questions.[11] This probing action "prepare[s] the physically described system to receive from nature a response to the particular question posed by the probing action" (Stapp, 2010: 6). The "Quantum Zeno Effect" is posited as the state of fixing on a particular question that is repeating the specific observation until nature replies with an answer. This is Process 3. This follows physicist Dirac (1930) who posits the answer by Nature is not known in advance and has an element of randomness to it. Process 2 is posited as the everyday, ongoing otherwise deterministic flows explained by the Schrödinger equations. These equations portray the movement of the probability cloud through time. Here, following Schrödinger, each state of the system reflects quantum possibilities, $|\Psi>$, potentialities or tendencies that are subject to collapse by observation. As Schwartz *et al.* (2004: 8) say, "process 2 generates a *cloud of possible worlds*, instead of the one world we actually experience" (emphasis in original). Within this evolving "smear" or "cloud of possibilities" we cannot at the same time precisely define location *and* momentum. Thus from each question posed to nature, which is said to be a conscious intention based on the abstract ego's own internal motivations and needs, some response will return which increases our overall knowledge. Each of these gains in knowledge becomes part of the larger pool of the knowable. Each question posed to nature already is a collapse of the wave function of possibilities to one in particular that is presented to nature. It is this question asking that provides the basis of new knowledge gains which become the basis of yet newer questions posed and ever newer bases of knowledge. Said in another way, the observer poses a question to nature in the form of either "yes" or "no" possible answers. Nature replies (Dirac's "a choice on the part of nature") by confirming either "yes" or "no." If "yes," then the observer experiences the effect that he was suggesting in the question posed. If "no," s/he does not have that experience.

In this three-step process of quantum consciousness, "the agent must first choose a probing action that he intends to perform, then participate in the execution of that action, and finally experience the resulting outcome" (Stapp, 2010: 6). "Probing actions," questions posed to nature cannot be predicted in advance, for they are the effect of agents with distinct intentions. Probing actions by an ab-

10. Schwartz *et al.* (2004:12) do note that these initial choices are "free" in the sense that current laws of physics do not show otherwise, but particular intentions are related to the past of the person.

11. Lacan (1977:313) has also suggested that a basic question is posed to the unconscious Other, "che vuoi?," what does it want from me? For Lacan, the Other responds by offering possible objects of desire that overcome lack.

stract ego produce two collapses of the wave function: the first, by posing a particular question in a "yes" or "no" possible answer format; the second by nature's choice—it chooses one or the other by principles or law unbeknownst to the prober. This choice on the part of nature is subject to quantum statistical laws. We have control in the question we pose, but not the answer. Process 1 is preparatory; it reduces all possibilities inhering in quantum states to a particular question. The answer can be either "yes" or "no." Yes "is associated with a specified possible experience that might occur in the stream of consciousness of the probing agent" (Ibid., 9). "No," no experience in the stream of consciousness. Unknown is which question is posed and what experiences will be associated with it in the stream of consciousness (ibid.). Nature's choice collapses the two possibilities into a "yes" or "no"; if "yes," the prober will have the associated experience in her/his stream of consciousness; if "no," no experience. If "yes" and with the associated experience in the stream of consciousness, motor activity is initiated in the form of "templates for action"—particular motor activity to achieve ends sort (Schwartz *et al.*, 2004). Thus here, the classical physical system kicks in to engage actual motor behavior.

Stapp also provides thoughts on how memory is stored. He dismisses answers provided by Newtonian classical physics that it is stored in brain traces. Rather, since in quantum mechanics it is the mental realm that acts on the physical realm (brain), then memory must be stored in the mental realm (ibid., 16). Although he stipulates a mental realm and mental structure, he does not precisely specify its wherewithal. In our view, it lends itself to a quantum holographic extension in showing that the mental realm is about quantum holograms, interference patterns, carriers of information that are stored not in the brain, but in the zero point field and enclosing boundary surface areas.

Stapp (2007: 24) has also posited a Process 0 by which he means the valuations and conscious reasons provided by the observer, the probing agent. He acknowledges that a "causal gap" exists in a more complete explanation from physicists as to the fuller nature of Process 0.

Finally, Stapp recognizes quantum retro-causality. Nature's choice reduces all possibilities (yes? no?) to one actuality. The other potentiality is retro-causally eliminated. That is, all information associated with the other possibility (either yes, or no) disappears. Thus quantum wave collapses "also eliminate from the records of the past all traces of properties that led to the possibilities that were eliminated by Nature's choice." Further, "the surviving records of the physical processes leading up to the collapse event exhibit only those parts of the past that lead up to what actually did happen" (Stapp, 2013a). This statement is in accord with the original "thought experiment" by Wheeler in 1978 referred to as the "delayed choice experiment." It follows the double-slit experiments first outlined by Young that established that an entity is both particle and wave (see chapter 2). Recent research has confirmed Wheeler's thought experiment. "Quantum erasure" and "delayed choice quantum erasure" experiments (see Kim *et al.*, 2000) show that what one has measured/observed (now seen as data) at one time, could disappear (quantum erasure) when we subsequently remove the measuring device or observation.[12] Or, said in another way, destroy the data (particles) and we now observe only waves and interference patterns. Thus Stapp's point is that once nature "chooses," all traces of the wave and interference pattern disappear; what we have instead is data, or information along with a history leading up to only this possibility. Information leading up to the other possibility disappears as if it had never existed. In other words, we perceive a linear, smooth flow of a stream of consciousness.

Both Penrose-Hameroff's and Stapp's models provide great insight into a quantum consciousness. We shall argue, what is also necessary is a holography addition. They also do not provide a more social model, that is, a perspective that would integrate inter- and intra-subjective components of agency. To this possibility we now turn.

Quantized QD

In previous work (Milovanovic, 2011, 2013a, 2013b) we offered initial approximations for a quantized inter- and intra-subjectively constituted subject. We began with the work of Jacques Lacan, a psychoanalyst, whose insights on the nature of the subject, expresses perhaps one of the most comprehensive undertakings the field has witnessed. He provides several models of the subject, some more static (schema L), some more dynamic (schema R, schema L, graphs of desire, Borromean Knots, le synthome; see Ragland and Milovanovic, 2004). His model is driven by a Hegelian notion of "lack" underpinning of the desiring subject. It is the primordial lack experienced by the infant in its introduction to the world of language and symbols that dooms it to forever being activated by a desire to overcome lack. This is a homeostatic ontology.[13] The subject is constituted by its ongoing efforts to overcome lack but whose responses are shaped in the field of language and its inescapable effects. Accordingly, he uses the terms "speaking" ("parlêtre"), or "the speaking sub-

12. See chapter 2, note 3 for illustrative web videos.

13. This is a tension reduction model that tends toward equilibrium conditions. Alternatively, dynamic systems theory suggests the cosmos is in far-from-equilibrium condi-

ject" ("l'être parlant"). Language, in his view, speaks the subject, not the other way around as conventionally understood. The "I" that takes up residence in discourse in order to speak is a product of complex feedback loops, anticipatory dimensions, and retrocauality ("retrograde"), particularly well portrayed in his "graphs of desire" and in schema R (see Lacan, 1977). His model, much like Freud's before him, is based on an oedipal structure, which is localized in the unconscious with real effects. Lacan's model is an inter- and intra-subjectively constituted diagram which portrays the complex ongoing relations with which the subject is engaged. It was Deleuze and Guattari's classic polemic, *Anti-Oedipus* (1972) against the Freudian underlying model based on lack and the Oedipus that paved the way for an alternative, de-oedipalized construction. Nevertheless, they often pay homage to the "master."

Lacan's schema R is particularly instructive for us in providing the take-off point for developing an alternative de-oedipalized subject based on an alternative ontology to lack and the Oedipus. But it is necessary to integrate quantum holographic principles to make this model more viable. Along with this, Bergson, Deleuze and Guattari, and Deleuze on his own, offer key suggestive philosophical works as to alternative possibilities. The works of Leibniz, Spinoza, and Whitehead also offer key insights. To this end, in the following section we further our work in offering closer approximations to the development of the subject, what we refer to as schema QD. Bergson's *Matter and Memory*, originally published in 1896 which inspired Deleuze's late two-volume work on cinema theory, *Cinema 1: The Movement Image*, *Cinema 2: The Time Image* are two key statements useful in developing a quantum holographic approach particularly describing the wherewithal of the materialization of perception-images and the subject.

1. Components of Schema QD

Figure 2 diagrams schema QD, as adapted from schema R. This portrays how "réalité," or in our adaptation, "perception-images" are created. It explains how they are instantiated from the plurality of possibilities that the quantum wave reflects. The relational configuration of components represents a QD matrix in-process, $|QD_i\rangle$. We first note the key components. We review a two-dimensional model, and then move to a more dynamic, three-dimensional model. We begin with the four components that are corners to

---

tions. Those in edgework criminology point out that people seek out excitement, seek tension, due to its activation of adrenaline rushes.

a central piece which Lacan (1977: 334) argued is but a flattened Möbius band.[14]

Figure 2. Schema QD

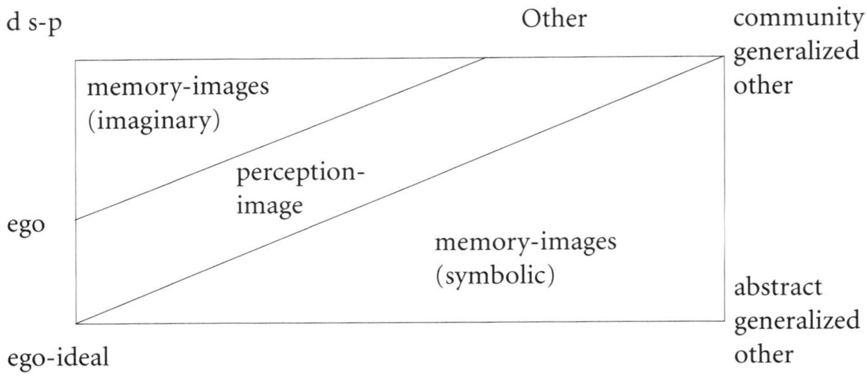

Components of Schema QD

Components of the central Möbius band
(each symbolized as a basis state, i.e., $|ego_i>$):

Ego: imaginary construction of a unified self through the eyes of the immediate Other.

Ego-ideal: identification with desirable traits one has internalized.

Other: the person either real or imaginary with whom one is in face-to-face interaction.[15]

Community generalized other:[16] some larger but more immediate significant group; "concrete generalized other" (Fraser, 1986); subcultural normative order.

14. A Möbius band is constructed by taking the central quadrilateral, invoking a half twist, then gluing the ends. Here, ego is connected with the community generalized other, the ego-ideal with the Other. The cross-cap in figure 3 is composed of an infinite number of Möbius bands, or, we shall see, "cuts" of the subject, each providing a momentary instantiated perception-image and an "I."

15. The Other ranges from an infinite, uncategorizable Other (Levinas, 1969, 1998) to more socially constituted, such as those appearing in the literature, summarized by Denzin (2007: 92) as: "emotional associations," acquaintances, strangers, we-relationships, contemporaries, peers, "situation specific," "intimate significant other," etc.

16. A generalized other, following Mead (1967), is some standard of a group which one assumes before acting.

Perception-image: unified image; bloc of space time; gestalt; réalité (Lacan, 1977); cut/bloc of spacetime; spacetime constrained holograms, or logons (Gabor, 1946).

Other components:

Discursive subject-positions (d s-p):[17] master categories, identities and indentificatory categories.[18]

Abstract generalized other: normative order; societal-wide dominant norms.

$|I_i>$: the speaking subject with its multiple possible forms of expression (i).[19]

Memory-images (imaginary): spectra; frequency interference patterns encoding information; "shining points" (Bergson), i.e., constellations/association of images.

Memory-images (symbolic): holographic lexicon; signifiers; expressive forms; including master signifiers (Lacan, 1977), "order words" (Deleuze and Guattari, 1987); icons.

In our refinement of Lacan we have three "others": the immediate face-to-face Other with whom we find ourselves in a reciprocal affirming attitude; the abstract generalized other representing societal normative structures; and an in-between community generalized other that reflects group standards, or subcultural normative order. The ego, or imaginary construction of self and its ideal are always interconnected with these three others in its construction of reality (réalité), or perception-images.[20] The ego—Other interaction is the inter-subjective (face-to-face) relation in which two interlocutors take up the position of the other. An intra-subjective component, the ego-ideal, conceived by Lacan as I(0), is always the backdrop ideal standard for the ego that engages in interaction with an Other. This in the wider context of a group and its standards/norms (com-

17. We make a distinction to earlier statements (Milovanovic, 2011, 2013) between an "I" and a structured d s-p.

18. This goes beyond "social role," in suggesting that each position provides an attached structured discourse and a locus of characteristically associated words (signifiers). Consider: traditional criminal justice identities—victim, offender, prosecutor, defense counselor, juror, juridic subject, etc.; identifactory categories—race, gender, religion, class, ethnicity, and emergent roles associated with, for example, LGBTQ; and other societally recognized forms. When applied to oneself (one's selves), this is much like the "me" of Mead's (1967) "I-me" constituted self.

19. The "I" has affinities with Mead's (1967) spontaneous, indeterminate, and unpredictable "I" that reacts to the "me" in providing actualized expressive form.

20. For more dynamic versions depicting these interactions, instructive is Lacan's (1977) "graphs of desire."

munity generalized other) and a society as a whole (abstract generalized other) that are contributing shaping factors of an ethical subject. Each of these (community and abstract generalized other) we conceptualize, following Bradley's (2010, 2011) lead, are established frequency patterns within the socius and within bounded regions (niches) to which conforming members form resonance.

Together, the inter- and intra-subjective components can be conceptualized as describing a phase space. This phase space would, then, witness recurring forms of attractors. A more molar configuration would be more along the lines of a point attractor, an agent that resists change, a becoming-of-the-same; a more molecular configuration is more in the form of a torus attractor that suggest an orderly disorder, a perpetual becoming-other, more along the lines of a dissipative structure[21] in chaos theory. In other words, each set of interacting dyads tend toward a particular highlighted distribution forming a distinct matrix.

This, we will show below, can be topologically diagramed on a unique sphere-looking object called a cross-cap. In actuality, it incorporates a Mobius band exhibited by a method of cutting into it following a figure-8. We will see how the flat Mobius band in figure 2 can be connected on its sides with a twist (see figure 3c). All four of the components comprising the corners of the Möbius meet along a vertical line (see figure 3a) which expresses the various possible instantiations of an "I." This matrix, which the human being is, can be described quantumly as $|QD_I >$. In other words, "i" suggests that the agent is a "cloud of possibilities" and its instantiations are more molar or more molecular in form. Said in another way, this matrix is a set of relationships, each defined and given form in context, each interacting with the others. It is not static, but is in continuous variation. At any moment, the matrix of interconnected components produces a "cut" of the subject, a "bloc of spacetime" which is the generator of a particular signature wave enfolding this configuration. One of its manifestations is the speaking subject, an "I" who takes up residence in structurally shaped discursive subject positions and particular forms of discourse in order to speak and construct reality. Consider, for example, a lawyer who takes up the position of a prosecutor and the circumscribed discourse s/he must employ. The speaking subject, or Lacan's speaking/speaking being, can be represented in quantum symbols as $|I_I >$. The subscript, small "i," refers to the possible expressions of the speaking subject, "I." The speaking subject, a person who inserts her/himself within a variety of narrative constructions as an "I," is but an instantiation of a cloud of possibilities. Her/his distinct "I"

21. The idea of a dissipative structure suggests that stabilities are constantly becoming destabilized and reassembled. Small inputs to these "structures" can have disproportional effects. Change, mutation, and metamorphosis are essential defining characteristics.

is always in-process, a becoming, instantiated within a context by inter- and intra-subjective components in the form of interacting dyads represented in schema QD. This suggests an inherently ethically-driven subject.

Unlike Lacan's model of desire based on lack, we substitute: Deleuze and Guattari's notion of "production," an ever-expanding search for actualizations of virtual potentialities (becoming rather than being), more along the lines of Bergson's "élan vital"; Nietzsche's will to power and affirmative action; and Spinoza's joyful encounters. Much revolves around Spinoza's ontological claim concerning "what a body can do" and its ability to cultivate "active affections," that lead to increased power to act. Apposed are "passive affections" that relate only to the capacity to feel, suffer, and diminished empowerment to act (see Hardt's insightful discussion, 1993: 91–111). We will return to further analysis in Chapter 6 on transformative justice.

2. Dynamics of Schema QD

In a good part of everyday interactions most scholars recognize that habit and passivity account for most responses. In Stapp's explanation, these can be explained by the Schrödinger wave equations of motion—how probability clouds in movement bear probability distributions we call "particles" or momentary instantiations of reality. It is some disturbance to routine (i.e., habitual everyday activity) that mobilizes creative channels and an alternative collapse of the wave function. Novel situations initiate active recognition (Bergson, 2002).[22] Thus in our more habitual activities, more dominant, conventional perception-images are instantiated, repeated and reinforced. These are interference patterns, or spectra. And in repetition there is a match between more stable spectra and situationally created logons. Here we follow Bradley's discussion of classical holographic principles indicating that these perception-images that retain a central place in our social field are spectra or interference patterns that have attained a degree of stability and resistance to change. They are attractors within a field of perception-images enfolding information. They represent quantum waves that reflect the potentialities embedded. Hence, Bradley's quantum holographic principles are implicated whereby subjects in interaction construct quanta of information, logons, in their ongoing activities. This information is but a cut of the possibilities inhering in the cloud of

22. Wendt, recall, has referred to an active monad, passive monad, and those that are more like objects. The latter being so because they are completely withdrawn from information. They are not part of the "social wave function."

potentialities. Drawing, again from Bradley (2011), it is pattern-matching, in our adoption, matching spectra (geometric encoded information, or holograms) representing the community generalized other, the abstract generalized other, and ego-ideal with everyday situationally-constructed logons. These are brought within sharpened focus in direct face-to-face interaction (ego—Other). It is this process that is behind the instantiations of the ethical subject. Let us restate this.

Consider Bergson's (2002: 36–46) classic statement of how images take form. Novelty is experienced as a question posed to the percipient (ibid., 46) with which we have previously seen agrees with Stapp's Process 1.[23] An object (entity, situation, some phenomena) is vibratory; it reflects frequency patterns. All entities emit and remit frequency waves representing enfolded information. A percipient receives these frequency waves, and in habitual mode, activates sensorimotor mechanisms that lead to more immediate unreflective responses. This is passive recognition.

However, in novel situations, an "encounter," an immediate response is suspended,[24] a "zone of indetermination" opens. It is at this moment that Bergson's (2002: 104–106) "circuit diagram"[25] suggests moments in which virtual memory images are accessed by different "leaps" until one appears as representative, a perception-image.[26] Thus a "doubling" (ibid., 101) of the external image takes place. As Bogue (2003b: 112–113) clarifies, "as we pay closer attention to the object [or encounter], we summon up memory-images from

---

23. It would also agree with Lacan (1977) who points out in his graphs of desire, that the "speaking" poses a question to the unconscious Other, "What does it want of me?"

24. Consider Lefebvre (2008:181), "a circuit can be established only by virtue of an encounter that inhibits or arrests movement. This encounter is, in the plainest sense, something we do not know what to do with. Rather than evoke a ready response, it initiates a prolonged, experimental search."

25. Bergson (2002: 104) tells us, "[R]eflective perception is a circuit, in which all the elements, including the perceived object itself, hold each other in a state of mutual tension as in an electric circuit so that no disturbance starting from the object can stop on its way and remain in the depths of the mind: it must always find its way back to the object from where it proceeds."

26. In the Penrose-Hameroff model, a rift in the spacetime fabric is engendered producing sheets which represent competing images of "reality." By way of "objective reduction," that is a threshold value of focused energy, one sheet comes to dominate, and the other(s) disappear. The spacetime fabric is momentarily stabilized. There is not much room, outside the effects of some "hidden variables" and the noncomputability thesis to account for how one image is instantiated over the others. In our model, we are suggesting the important contribution of the signature wave of the percipient which is momentarily instantiated in matrix form representing the dynamics of schema QD.

broader and more distant past contexts, each wider context encompassing the narrower." For Bergson (ibid., 101) "an appeal is made to the deeper and more distant regions of memory, until other details that are already known come to project themselves upon those details that remain unperceived." In other words, these memories are projected onto the external object or encounter filling it in with possible content providing several possible versions of a completed picture, until one is "chosen." In our adaptation, these renditions are in superposed, quantum coherent states until the wave function collapses producing one instantiation.

Bergson then argues that the light emitted by the object is returned along the same pathway to the object, illuminating it (2002: 45). In his words, "everything thus happens for us as though we reflected back to the surface the light which emanates from them, the light which, had it passed on unopposed, would never have been revealed." He warns us that this is not like a photograph of an object which is then processed in the brain "by some unknown chemical and psychical process of elaboration" (Ibid., 38), but rather, "if photography there be, [it] is already taken, already developed in the very heart of things and all the points of space ..."[27] Thus, when a zone of indetermination is activated, virtual memory images are recalled seeking a match with the present emitting object or encounter. A "chosen" image, encoded within a distinct signature wave of the percipient, is then projected out along the same pathway to the location of the emitting object/encounter, "illuminating" it.

We have support here in Robbins's (2006) suggestion that the brain, building on Bergson, can be conceptualized as a decoder, a "modulated frequency wave" traversing a field. A surface or field that encodes information holographically can be accessed, according to Robbins by way of changing frequency values which reproduce a particular three-dimensional object (ibid., 372–373). "The brain," he continues (ibid., 373), "is a decoder; it is the reconstructive wave that is unpacking the 'code' in the holographic field." In this view, the brain is a "modulated reconstructive wave" in continuous variation.[28]

---

27. Robbins (2006: 367) commenting on this point has argued that "Bergson was envisioning the matter-field as a dynamic interference pattern—a holographic field...," or the ZPF.

28. Robbins (2006: 373) states: "The information in the field-wide, dynamic interference pattern the wave is picking out and specifying would be precisely the invariance laws or invariance structure described by Gibson." He also notes that "abstract space/time is a projection frame for our thought, derived from the necessity for practical action." The initiation of these modulated frequency waves, he (ibid., 390) tells us, is based on everyday cues. In our view, lines of flight, or affordances, could be the cues that initiate the process, which

Thus, in adaptation, the unique signature wave of the percipient, the result of a particular matrix formed in context of its interacting relations, is this modulated reconstructive wave.[29] What we have added is the underlying development leading to the emitting signature wave. Let us see how this can be reconceptualized further by employing phase conjugation and Lacan's topological construction of the cross-cap portraying the dynamic version of schema R, the latter of which will be reformulated into a quantized schema QD.

Phase Conjugation

Bohm (1983), Pribram (2006: 43), and Mitchell and Staretz (2011: 5) have noted that if it were not for the lenses (transducers) that our eyes and other sensory apparatuses represent our cosmos would appear as a hologram, a two-dimensional world, much like the recording screen in the two-slit experiments. Let us draw from the notion of phase conjugation to bring quantum holography to support and extend Bergson's profound discovery. And remember, Bergson's classic work was published in 1896, a good 25 to 30 years before the main principles of quantum mechanics were discovered. Schempp and Marcer's groundbreaking work along with the synthesis offered by Mitchell are quite compatible with Bergson's statements in *Matter and Memory* concerning how some object, ontologically defined by Bergson as existing in a vibratory state, undergoes an instantiation. For Bergson, all objects/entities emit and remit waves. In phase conjugation dynamics (Pcar), an incoming wave of an object reflecting its potentialities is met by an outgoing wave, sometimes referred to as a "reconstructive wave," by the recipient producing a "standing wave," an interference pattern; in short, an

because of schema QD matrix being in continuous variation, and the uncertainty principle at work, nonlinear end states result. See also, Robbins, 2007; 2010: 516–517; 2012: 24–33, 56–68.

29. Here we are in disagreement with Cramer (1986) who argues that consciousness/agency is not a necessary ingredient ("TI does not need to talk about brain states, consciousness, or observers (rational or otherwise")) in the construction of the perception created in the real. It is, in our view, the distinct signature wave of the moment, a result of a QD matrix collapse of the wave function, returning to the emitter that provides a more narrow range of possible instantiations at the source (initial emitter). We are in agreement with Cramer, and follow Bergson, that reality is finalized at the source (emitter), not in the absorber (brain) and that a more restrictive stochastic choice is made there. Cramer, consistent with quantum atemporality, argues no "real" time appears to take place in the process. All is perceived as seamless in the real.

informational communication channel opens.[30] This outgoing, reconstructive wave we see as a signature wave reflecting the dynamics of schema QD and its momentary instantiations in matrix form reflecting the intensities of each of the components in context. From this momentary instantiated matrix reflecting schema QD, a speaking subject, an "I" emerges. It is this dynamic that produces a distinct signature wave encoding the matrix that in its emission also encodes the collapsed incoming wave. The incoming wave, recall, was subject to the dynamics of Bergson's circuit diagram whereby memory images were accessed for an appropriate fit. A collapse of the wave function to one image, now a recollection image, becomes intertwined or encoded within the signature wave of the percipient that is taking form. The finalized signature wave is emitted outward back to the source of the incoming wave creating a three dimensional image "out there" which we now see as reality.

Let us say this in another way. The percipient, exposed to novel incoming vibratory waves, quantum waves holographically enfolded with information, mobilizes Bergson's zone of indetermination. Within this contemplative zone, or momentary suspension of action, Bergson posits via his circuit diagram, how virtual images ("recollection images") are created from memory ("pure memory") that are sought to match the present incoming image. Deleuze (1989), building on Bergson, has demonstrated how certain cinema theorist's work (e.g., Robbe-Grillet, Resnais) are consistent with this process. In their work they suggest either "sheets of the present" or "sheets of the past" are created in viewing cinema. Deleuze informs us that this is an exemplary case of how zones of indetermination are initiated, developments which are consistent with late modernity's form of image making he calls the "crystalline regime of signs." This regime is characterized as engendering zones of indetermination in its very organization.[31] These "sheets" represent alternative side-by-side existing possibilities of "what happened?"; or reconceptualized, this is a state of quantum superposition, or quantum coherence. For Bergson (2002: 241), "we believe that we can discover thousands of different planes of consciousness, a thousand integral and yet diverse repetitions of the whole of experience through which we have lived." Deleuze (1986, 1989; for a precise, accessible explication, see Bogue, 2003: 35–39), agreeing with Bergson, in his two-volume cinema studies, has also shown how this zone of indetermi-

30. Robbins (2006: 373) has said: "the brain is a decoder; it is the reconstructive wave that is unpacking the 'code' in the holographic field."

31. As opposed to the "organic regime" whose central organization revolves around continuities, linearities, and order.

nation is the locus of affection-images, a qualitative dimension, that remain within the body.[32] Affection-images defined by Deleuze (1986: 66) are "a sort of motor tendency on a sensitive nerve," or sensations and bodily feelings. Perceptions are created in conjunction with bodily sensations.[33] Thus affections are in the body,[34] perceptions in the emitting object with which we are interacting. Denzin (2007: 3, 48, 93) has argued that emotions are always constituted inter-subjectively with always a present, implied, or imagined Other. Action-images are sensorimotor schemas that are always already implicit in the perception-images created. They produce (Deleuze, 1986: 65; see also Bogue, 2003a: 35) an "incurvation of the universe, from which arise simultaneously the virtual action of things on us and our possible action on things." In other words, each percipient is the locus around which the universe curves according to our expectations and anticipations (ibid.; see also Bogue, 2003a: 35–36). For Bergson and Deleuze, the subject is a composite of its perception-images, affection-images, and action-images. The brain, then, according to Bergson and Deleuze can be perceived only as an interval, a switchboard, a center of indetermination, a specialized image amongst the universe of images. Deleuze (1986: 63) also suggests that these specialized images, that we are, were primordially constituted after the Big Bang during a cooling down of the plane of immanence which produced initial "screens obstructing the diffusion of light" and producing "the first outlines of solids or rigid and geometric bodies." These outlines are vibratory in nature. These "outlines," we have already seen, are a response to the wave function collapsing in uniform ways.

The eventual momentarily instantiated perception-image is a situationally created logon, a result of the instantiated matrix composed of the interaction amongst various components of schema QD. At the same time as the perception-image is created, so too the subject. They are simultaneous events giving the appearance of a centered, determining subject. Schema QD and the matrices it forms are the origins of a unique signature wave of the percipient.

32. We also depict, below, that its loci is the "disc" that is the center of the cross-cap.

33. The importance of affection is most often missed in contemporary criminology. The notion of "edgework", coined by Steve Lyng (2005) has brought back the emotionality dimension as a critical component of subjectivity. See also Katz (1999) and Sartre (2002). Denzin's (2007) and Katz's (1999) work, unfortunately, remain little noted in the literature in contemporary criminology.

34. Hameroff and Penrose (1998) have called affections qualia. They are encoded at the Planck scale as geometrically organized "spin networks."

An "I" that takes up residence within various discourses as a speaking subject, reflects the momentary configured matrix created and the distinct signature wave that is its reflection. The "I" has a degree of stability, making for the identity that all confer on it, but can be actualized in many ways. Thus the outgoing wave interferes with the incoming wave producing an interference pattern, a hologram. The recipient, via her/his sensory apparatus, or transducers, makes use of forward Fourier transforms to construct wave forms, and a returning signature wave. The percipient's signature wave, collapsed in a particular configured matrix form, both is emitted outward along the path of the object bearing wave, and inward activating the motor apparatus of the body. Through a reverse Fourier transform the object is constituted "out there" from where it came; not in the brain. At this point motor action has begun, but full consciousness is only attained on the return of the perception-image. This is consistent with much research that shows that we often initiate motor action before we are fully conscious of what we are doing.

Consciousness, Bergson tells us, is not *of* something as the phenomenologists insist, but *in* something. On every occasion of instantiation the recipient encodes information about her/himself and transmits this in wave form to the objects it constructs "out there." In this view, we are inherently part of the cosmos. With the notion of non-locality, we are intricately connected with others, the community, the state, and the cosmos itself. Herein is constituted a spiritual connection that more and more of those in quantum holography research are acknowledging.[35] We already witness the fact that many of the founding figures of quantum theory in the 1920s did indeed have some involvement with a spiritual dimension (see chapter 2). There is, then, no intrinsic separation of the "I" from the Other, the Other from the "I."

Let us clarify a bit more. An emitting entity's frequency wave is received by a percipient, whose sensory apparatus acts as a transducer producing forward and reverse Fourier transforms. In the first stage, we see this incoming frequency wave, in the active recognition mode, generating a zone of indetermination, perhaps, following Penrose and Hameroff, a split in the fabric of space time. Within this zone, we see Bergson's circuit diagram at work producing various competing virtual images that are being matched with the incoming information. We also integrate Mitchell and Staretz (2011: 7) who note that at this moment the percipient is activating memory images located in the ZPF, not in the brain, by way of her/his unique signature wave. A resonance condition is established between the brain's activation of microtubules

35. See the summary of empirical analysis by Radin (2013).

and memory stores in the ZPF (ibid). Memory images are spectra, holographically encoded information stored as frequency interference patterns. But unlike Pribram and Bradley who argues they are Fourier descriptors stored in the brain, be it non-locally, Mitchell, following Laszlo, argue that all information storage remains non-locally dispersed in the ZPF and can be activated by the appropriate frequency that the percipient's unique signature wave provides.

With resonance, an informational channel materializes during which, and consistent with Bergson's notion of a zone of indetermination, recollection images are created via the circuit diagram. These remain in a quantum coherent state, following Penrose and Hameroff, perhaps as sheets, each of which depicts competing versions of reality. Following Mitchell and Staratz (2011: 14), "evaluation" can then take place of the respective images and meaning assigned (ibid.). Following Hameroff and Penrose, we conjecture that with the predominance of meaning attached to one possibility rather than another, an intensity level is reached at which point an "objective reduction" (collapse) of the wave function ensures. The superposed, or quantum coherent state, collapses into one image, a virtual image that becomes intertwined with the percipient's emitting signature wave. And the initial attention transforms into intention. Our point is further supported by Delanda's (2006: 20–25) notion that each entity or assemblage has differential capacities to effect and be affected, and thus has critical thresholds at which action is triggered. Thus similar stimuli may have dissimilar effects.[36] Evaluations, then, are nuanced. Mitchell and Staretz's notion of an evaluation stage implicates the ego-ideal, the concrete generalized other, and the abstract generalized other and their interactive work in shaping of the eventual perception-image. This can be located in figure 2 as a flattened Möbius band.

The eventual perception-image begins to take form. Frequency waves reflecting its informational basis are emitted outward along the same pathway as the incoming frequency wave, but also radiating throughout the body, which prepares it in an anticipatory manner for action. This explains the seeming paradox that action begins first and consciousness of the action follows up to a half second later, as in the Libet *et al.'s* (1970, 1983) studies. This can be connected with the opening of the zone of indetermination and the processes taking place (e.g., Bergson's circuit diagram). The second stage is where the percipient's outward going wave is delivered back to the emitting object producing a three-dimensional image. This is also the basis, in our view, of the

36. Delanda thus suggests that nonlinearity is the rule, and what constitutes the threshold value are unique "triggers" to which an entity responds. He suggests we use the term "catalysis" as a form of nonlinear causation.

retrocausal effects that perplex many quantum holographic theorists since Libet *et al.*'s experiments. This image then returns to the percipient to complete the perception-image and becomes the basis of distinct motoric action.[37] Due to retrocausal effects, and the elimination of competing explanations, an appearance exists of a linear development from stimulus to action.

Topological Model: Cross-Cap and "Figure 8" Cuts

Let us explain the dynamics once more, this time with the use of topology.[38] Consider Lacan's presentation of the more dynamic version of schema R which he sees as portrayed by the topological diagram referred to as a cross-cap (see figure 3).[39] The cross-cap can be depicted in three-dimensional space, but more naturally inhabits a four-dimensional space, since there are no boundaries to be crossed. Take for example a knot in 3-dimensional space. In 4-dimensional space it becomes unknotted. The cross-cap is unique in that it contains properties beyond a sphere; specifically, that an infinite number of Möbius bands can be inscribed within it. Lacan uses the cross cap to more accurately portray how a Möbius band[40] constitutes a perception-image, which he referred to as "réalité," a gestalt. This is also Gabor's spacetime constrained logon. In figure 3a we note a closed cross-cap and its unique features particularly with how an "inside" finds itself connected with an "outside" without any boundaries

37. Penrose and Hameroff (2011: 17) and Hameroff (2007a: 1040; 2012) have also noted this retrocausality. Even as a visual image is taking form quantum information is being referred back in time to the beginning of this materialization, filling it with content. Cramer's (1986, 2001, 2006a, 2006b) transactional interpretation also provides some compatible explanations of retrocausality. In his stage 4, quantum information is passed forward and backward in time, "reinforcing the selected transaction repeatedly until the conserved quantities are transferred and the potential quantum event becomes real" (2006: 22). The collapse of the wave function is witnessed at the point where an instantiation takes place, seen as the "real," from the possibilities.

38. For a fuller visual expose of topology (stick with the images not the broken English), see: http://web1.kcn.jp/hp28ah77/index.html#chapters. For the cross-cap, see: http://web1.kcn.jp/hp28ah77/us27g_crsp.htm. See also Carter (1973) and Barr (1964).

39. For a vivid depiction of the Mobius band and how it relates to the construction of the cross-cap, see Part 3, "Projective Geometry," section number 39, at the following site: http://web1.kcn.jp/hp28ah77/us27g_cros.htm.

40. Take a rectangle, twist it once and connect the ends. You have a Möbius band. Trace an ant walking on its surface and you'll note that it would not perceive any boundaries on its journey.

Figure 3. Cross-Cap and Schema QD

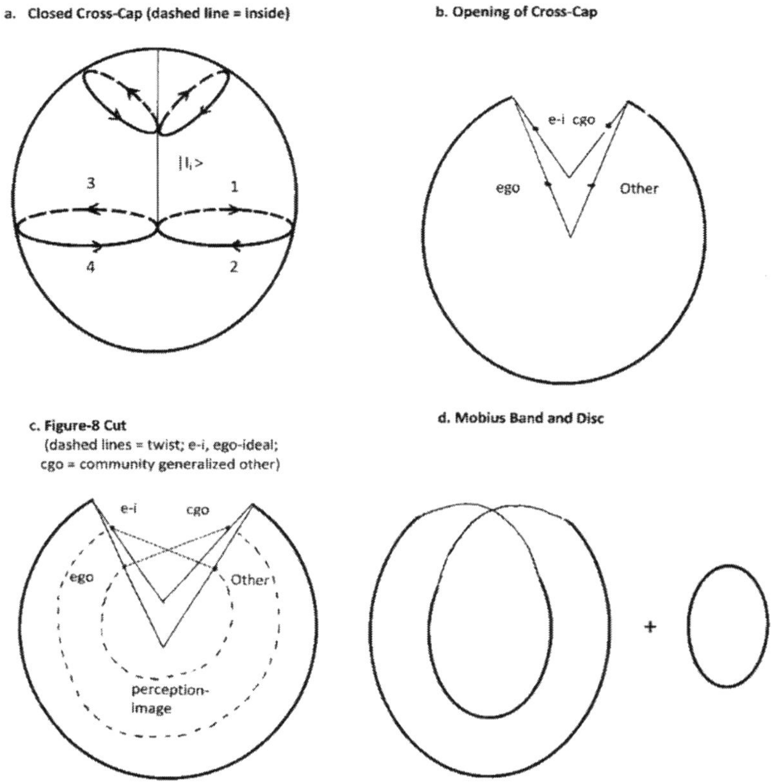

Source: Adapted from Granon-Lafont (1985) and Dor (1992).

crossed. An ant, for example that follows the pathway 1,2,3,4 will perceive no boundaries, going "inside" the cross cap and emerging on the "outside," only to go back "inside," etc. This crossing from the inside to the outside and back, also implicates affection-images which are retained in the body (Bergson, 2002; Bogue, 2003a: 37). As Bogue says, "perception takes place in the object, affection in the body"; and Bergson (2002: 57), "affection is thus that which, from the interior of our body, we mix with the image of exterior bodies."[41] All

41. Bergson (2002: 57) underscores the point: "just as external objects are perceived by me where they are, in themselves and not in me, so my affective states are experienced where they occur, that is, at a given point in my body."

perception-images are embodied with these internal affections. Said in another way, the internal eight, indicating how the inside is connected with the outside portrays how affection-images (in the body) are inescapably connected with the outside. Figure 3b situates the four corners of the Möbius band. Figure 3c portrays the "twist" that connects ego-ideal with the community generalized other, the ego-ideal with the Other, the ego with the community generalized other, and ego with the Other. The closed cross-gap indicates the continuous interacting dyads.

Figure 3c shows two cuts into the cross-cap. Lacan says the first moment opens up the cross-cap, exposing its constituent components; showing, for example, how the opposite corners of the Möbius band can reconnect. The second is a figure-8 cut (i.e., scissor cut following the borders of a Möbius band, see dotted lines, figures 3a,c). We must think mathematically and topologically. We are indicating how a figure-8 cut carves up topological space in a particular way with important effects. Thus the second cut depicts how the completed cut produces two objects (figure 3d): one, a Möbius band which reflects the instantiated perception-images and schema QD matrix; the other, a disc. For Lacan the disc represents objects of desire and its spiraling to a point, much like a seashell, the phallus that colors all language, thought, and action. For him, all is tainted by the law-of-the-father. We, rather, conceptualize it as: a loci of virtual recollection-images (RI) in a state of quantum coherence (in quantum language, written as a basis, $|RI_i>$), or superposed sheets each indicating a rendition of the "real"; affection-images mobilized in the encounter between percipient and others, more amorphous as we spiral down toward the singularity; and the singularity itself representing the portal for non-locality and "quantum tunneling." Topologically speaking, we conceptualize layers in this spiraling disc: the top surface layers are the competing virtual recollection-images; further down, a layer of what Bergson referred to as "pure memory images," or for us, spectra; then spiraling downward, a deeper area of affection-images; and the convergence on the singularity, the opening up to tunneling effects. All is patrolled by the abstract generalized other, a normative order, or relatively stabilized frequency patterns within the in-formation field to which agents become a-tuned. Master discursive subject-positions, or discursively constituted social roles in tune with the abstract generalized other[42] provide a loci of subjectification (see next section) from which a disciplined speaking subject, an "I" emerges. These are historically situated, political economic

42. In some cases, phase resonance exists between the abstract community generalized other, in others, discordance.

constructs.[43] Their expressive forms are shaped by the logic of abstract machines manifesting more molar or more molecular assemblages.

For Lacan, and for our revised schema QD, the subject is the figure-8 cut carved from the cross-cap. This is a bloc of spacetime, a cut of the subject; from its many possibilities and potentialities one momentarily appears, with an "I" that allows it to speak, and a distinct signature wave that encodes its distinct matrix at that moment. Between moment one (opening of the cross-cap) and moment two (completed figure-8 cut) we see "fledging selves" emerging (Massumi, 1992: 75–76), clouds of potentiality, but their expression shaped by disciplinary forces and "technologies of the self" (Foucault, 1997, 1994). Fledging selves could also be conceptualized as momentary superposed states of possible expression (quantum coherence). At the completion of the figure-8 cut we see what Massumi (ibid., 76) calls a "full-fledged self": "a full-fledged self only takes wing after the grid of value judgment has been successfully applied to the body, incorporeally transforming it into its assigned category." This is the collapse of the wave function. Possible expressions have decohered into one that becomes representative of the subject. We could also depict a moment before moment one, a time where spacetime movement is depicted by the Schrödinger quantum equations of motion reflecting probabilities within clouds of potentialities.

In this model, following Lacan, the vertical line in Figure 3a, is a "line of identification," the moment at which the various figure-8 cuts self-intersect and traverses the inside to the outside.[44] It expresses the potential forms of instantiations of an "I," or: $|I_i >$. It reflects the totality of the matrix instantiated at that moment. Thus perusing figures 3b and 3c we can envision that when the cross-cap is closed, all four corners of the möbius are connected with the help of the twist. This indicates the interconnected and mutually interpenetrating nature of these components. Along this self-intersecting line a particular "I" is instantiated, a speaking subject, with a degree of unpredictability as to the precise form it takes. The agent's unique signature wave, emitting constantly outward, reflects this instantiated matrix and the "I" that is its representative.

43. See, for example, J. Young's *Criminological Imagination* and his discussion on "othering," to which we also return in the next chapter. For Young, the way we perceive the other is centrally rooted in political economic conditions. We build on his notion of othering in showing that three distinct "others" can be conceptualized, and by doing so, we believe a better understanding of agency follows.

44. Lacan has also referred to this as a "line of intersection." In three dimensions it is so; in four, no intersections exist.

Quantum Holographic Model of Language and Speech Production

We are also in a position to make some remarks on a quantized version of Lacan's (1977) "graphs of desire." He developed four graphs ranging from a seemingly simpler to a more complex model with multiple feedback loops. These are diagrams of the various movements of the various components of the subject which shape speech production. He argues that there are two temporal dimensions at work: a forward anticipatory dimension, and a "retrograde," or reversal to the beginning of some narrative unfolding. For us, the latter implicates reverse causality, for the former, Gabor's (1946: 437) notion that logons include in their sine wave composition an "overlap of the future," an anticipatory dimension. Let us develop this more, but first, just a quick review of the nature of images.

The question of the wherewithal of images and the specialized image we call the agent is addressed by Deleuze in *Cinema 1* (1986: 63) building on Bergson's *Matter and Memory* as well as *Creative Evolution*. He begins with the earth when it was very hot, perhaps suggesting a point right after the primordial Big Bang, where the plane of immanence was a sea of energy in motion where only images circulated (ibid, 58). On this plane these movement-images are "flowing-matter," are loci of energy. The plane of immanence is composed fundamentally of light (ibid., 60). Thus in Deleuze's reading of Bergson, matter is light. These images, Deleuze continues, "are not yet bodies or rigid lines, but only lines or figures of light. Blocs of space-time are such figures. They are images in themselves. If they do not appear to anyone, that is to an eye, this is because light is not yet reflected or stopped ..." With the cooling down phase the "first opacities" took form, along with their abilities to interfere with the diffusion of light. "It is here," he tells us (ibid.), "that the first outlines of solids or rigid and geometric bodies would be formed." It is at this time, too, following Bergson, that special images, "living images," are formed which have the capacity to delay action, to mobilize a zone of indetermination, in which alternative courses of action (action-images) are considered. Here, the rays of light that reach the percipient, "instead of passing through those centers, will appear to be reflected and thus to indicate the outlines of the object which emits them" (Bergson, 2002: 17). It would be at this point, too, following Bergson's *Creative Evolution* (1989: 300–304) at which the three fundamental categories of language are formed in response to interacting with the oscillating, vibrational whole which is the cosmos (see also Deleuze, 1986: 59–60; Bogue, 2003a: 36–37): nouns, verbs, adjectives. Adjectives are qualities that begin to reappear with discernible degrees of differences.

Nouns appear when boundaries to recurring qualities take form, "a snapshot view of a transition" (ibid., 302). Verbs appear with repeating acts.

From this background, Deleuze further develops a taxonomy of signs reflecting variations in movement images, "luminosities," on the plane of immanence. Here, in the field of semiotics as opposed to linguistics, what C.S. Peirce referred to as a "phaneron" the image "is a fundamental appearing in a process of continual change" (Rodowick, 1997: 55). This field is composed of images of varied modulations/vibrations. From this, Deleuze deduces six kinds of movement images which together produce "signaletic material."[45] Opposed to this is the field of linguistics that imposes the structure of paradigm (word choices) and syntagm (syntax and codes, e.g., appropriate linear constructions of words into coherent sentences). Linguists inform us that we choose words (signifiers) from a vertical axis (paradigm) and follow some code in stringing them correctly in linear order (syntagm) to make sense. Thus in de Saussarian linguistics, the signifier (S) is an acoustic-image referring to its object, the signified (s). Say a tree, for example, and an image of a tree appears, thus S/s.

Marcer and Schempp have been instructive as to how icons, or signifiers, are activated. In their model explained in "Quantum Holography as a General Strategy for Modeling for Semantics" (1999: 463–465), they demonstrate how "reality" on the one side, as composed of holographic encoded objects or icons "maps to" the "holographic encoding of the arbitrary label for that object or icon" on the other. By the latter is meant a common language. The arbitrary label is a signifier. There is a two-way process: each holographically encoded object encodes a particular label for that object; and in a reverse direction, each holographically encoded label decodes "reality" by way of phase conjugation. In other words, the object image bearing wave and the non-object bearing wave[46] create interference patterns, resonance, during which "reality" is either coded or decoded. Each holographic encoding of an entity or icon is encoded by the arbitrary, but consensually understood, holographically encoded signifier that represents it. The reverse process reconstructs the holographically encoded information. The sensory apparatus, as frequency transducers, or "quantum holographic transducers," either decode in an incoming direction or encode in an emitting outward direction this standing wave. Transducers are also the source

45. This is the basis of Deleuze's (1986) first volume on the "movement image" and "organic regime of signs." The second volume (1989) brings up an alternative regime of signs, the "time-image," the "crystalline regime of signs," whose distinct capacity is to break from the linearity of the movement image, in creating social reality and engendering creative responses to incoming object bearing waves.

46. This is often referred to as the "reference wave."

of the signature wave of the percipient (Marcer and Schempp, 1998: 237).[47] They are responsible for creating virtual object images sent back to the emitting object creating a 3-dimensional image "out there" where it is experienced. Experiences labeled in a similar way become the basis of a resonating frequency interference pattern. More stabilized as spectra, these become the basis of common background assumptions, or a resonating conceptual space. For these authors (1998: 235), "languages are shared/common arbitrary labeling of objects and their properties that makes this implicit iconic phase conjugate labeling explicit."

In the absence of some immediate signal or emission located in space-time, Mitchell (2008: 9; Mitchell and Staretz, 2011: 13) has suggested that studies in "remote viewing"[48] have indicated that "clues" or simply an icon reflecting an object, residing in the percipient's "internal data base," can activate a phase conjugate process. In our conceptualization, this internal data base comprises memory-images (symbolic and imaginary) of the percipient, stored non-locally in the ZPF. These are more stabilized spectra, frequency interference patterns holographically encoding information. These would include what Bergson (2002: 171) refers to as "shining points" or "dominant memories ... round which the others form a vague nebulosity." Recollection-images, however, drawing from these pure memory images, are more like logons, or ongoing, momentarily constructed quanta of information.

Recent empirically-based word association studies by Bruza *et al.* (2008) in understanding the human quantum lexical structures,[49] the semantic structure of language, suggest that it is context that acts as an observation/measurement that collapses the wave function. When a context is ambiguous multiple possible senses are activated in a superpositioned state; that is, a state of quantum coherence (Bruza *et al.*, 13). With unambiguous contexts, the particular constellation of associated words in superposition, or in a quantum coherent state, collapses to one value (Bruza *et al.*, 2009a: 7), and all information about other possible senses are erased (ibid., 13). A context, then, acts as a cue[50] for

47. The source of "coherent, non-object reference signal beams" within the sensory apparatus, or transducers, are said to be the fine interconnecting structures amongst neural dendrites. Nuclear membranes and the brain cortices are hologram planes (Marcer and Schempp, 1998: 237). Gariaev *et al.* (2001) and his colleagues have made an argument that it is DNA that acts in this capacity.

48. For a comprehensive review of remote viewing studies, see Radin (2013).

49. Bruza *et al.* (2009a) have defined mental lexicon, as "the words that comprise a language, and its structure is defined here by the associative links that bind this vocabulary together."

50. See also Robbins (2006: 390) who states "cues" are what may evoke reconstructive waves. He seems to suggest icons will do which tend to have an invariant quality. See also

the collapse of the wave function.[51] And a word, in a particular context, can generate particular word associations (ibid., 7). We extend this to the work of Gibson (1966, 1982), who indicates contexts or fields offer "affordances," or possibilities,[52] via ambient light. Perception of these affordances are said to be "direct," not needing further conscious effort. Affordances in our reconceptualization are spectra encoding (holographic) information. Phase conjugation dynamics along with forward and reverse Fourier transforms explain how these are constitutive of the "reality" we perceive within our spacetime. It has also been suggested that a "niche" a living being often establishes within which to operate can be viewed as a constellation of affordances (Jenkins, 2008). Of course, without more, this classic form of holography suggests holographic reductionism. Needed, for a fuller nonlinear understanding is quantum holography to indicate how ongoing, changing reality is conceptualized at any moment in the form of logons (Bradley, 2010, 2011).

We see here how Lacan's work once again is a starting point for refinement in this process. Lacan tells us that normally we find signifieds sliding under signifiers until a punctuation point is attained, symbolized as s(O). Say the word (signifier) "bounce," for example and a constellation of associated signifieds make their appearance in a quantum coherent state. Thus: S/s-s-s-s-s ; or, expressed in quantum mathematical symbols, $|\text{Bounce}_i >$. Said in another way, during the various feedback loops diagramed in the "graphs of desire," a superposition is created with a number of competing possible signifieds for a particular signifier.[53] They are momentarily in a state of quantum coherence. What collapses the wave function to one particular signified is the combined effects of anticipatory and retrocausal factors that punctuate, collapse, all possibilities to one that becomes the instantiated signified attached in context to a particular signifier; for Lacan (1977), symbolized as the "punctuation points" ("point de capiton"), s(O). In more detail, in the anticipatory dimension recall Gabor's logons characterized as overlapping with other logons producing within each a part of the next, or in Gabor's words, "overlap of the future." This suggests Lacan's anticipatory dimension. Each new signifier that is taking form, to cite

---

Marcer and Schempp (1998, 1999; Mitchell, 2008; and above) and their discussion as to how signifiers ("arbitrary labels") map to and from signifieds by way of phase conjugation.

51. Bruza *et al.* (2009a: 26) insist that their model makes no larger claims as to a quantum holographic processing brain.

52. These are said to be invariant factors within an ecological space; objects within this space have a particular meaning.

53. Consider Bruza *et al.* (2008: 2); "it seems possible that word meanings are somehow entangled, with the different senses of words all being active in some way until we actively create meaning by collapsing what is potentially a massively entangled state."

Lacan, "represents the subject for another signifier," until some pause, at which point the retrograde is back to the beginning of the string of signifiers. This retrocausal effect produces meaning for the whole sentence uttered (S-S-S-S. / s), and provides distinct values for each of the signifiers only after the completion, the punctuation point, s(O).[54] Communication channels are opened to one's particular lexical storehouse (memory images) via phase conjugate dynamics (Marcer and Schempp, 1998, 1999). Signifieds remain in superposition, in quantum coherence, while the percipient is represented by a signifier embodied with a quantum wave of information, a unique signature wave, encoding the dissipative matrix of schema QD in a particular context. Backward time referral consists of the emerging dissipative matrix, in the mathematical symbols of quantum theory, $|QD_i>$, returning to the beginning of the sequence of signifiers infusing all as finished product. This is similar to Lacan's "retrograde" by which a return direction, initiated by some punctuation at the end of a constructed sequence of signifiers produces understanding of the whole sequence, all at once, or s(O). This is the collapse of the wave function, S/s. It is also at this point that a relatively stabilized matrix representing the various interacting components of schema QD is instantiated; in other words, this is the second collapse of the wave function. From the possible, $|QD_i>$, one momentarily expressed unitary subject appears. This provides meaning and an instantiated speaking subject, an apparently centered "I" guiding the process, and concurrently, a distinct signature wave emitted encoding this instantiation.

Recent theorists have suggested that the activation of signifiers with their associative meanings takes place in context. A context, for Bruza *et al.* (2009a: 9), may include a sentence within which a new signifier is at question, but also factors such "genders, cultural groupings, historical periods, etc." We can extend their notion of "context" to mean social groups, organizational settings, niches, etc., and also to distinct forms of discourse (legal, scientific, religious, street corner, trial court, etc.). Within each context, distinct cues, or, in our appropriation following Gibson, distinct affordances are ubiquitous. These cues often appear in the form of recurring, circulating signifiers within an ecological niche. Deleuze and Guattari (1987) have suggested that words are not neutral but have a performative function. They are "order words," or for Lacan, "master signifiers," that evoke action (action-images). Bruza *et al.* (ibid.) do

54. Take an example. If I say: "I would …", so far, no meaning, " … like to", so far no meaning, "drink milk." Notice that the complete meaning of "I would like to drink milk," or symbolized as S-S-S-S-S-S., only occurs at the punctuation point (here, the period) at which point we return to the beginning of the sentence grasping the whole all at once.

leave off their more quantum approach with a nod toward the possibility of a holographic human lexicon.[55] We vary from Bruza *et al.* and alternatively side with those advocating phase conjugation in the access of information. In this approach context indeed provides cues, but what are accessed and activated are spectra, interference patterns encoding holographic information that reside in the ZPF, not in the brain.[56] We also invoke the idea of the collapse of the wave function, since any informational bearing wave is also a cloud of possibilities from which eventually one distinct perception-image is instantiated. Or, alternatively, with momentary breaks of the otherwise smooth flows of more habitual consciousness (inattentive recognition), a zone of indetermination opens much like that suggested by Penrose and Hameroff's bubble in the spacetime fabric. It is within this moment that the percipient actively constructs recognition-images in attempts to understand a novel or ambiguous encounter. The superpositioned (quantum coherence) recognition-images collapse to one that becomes a virtual image that, along with its inclusion of the signature wave, is re-emitted back to the object producing a standing wave culminating in the 3-dimensional spacetime objects we see as reality.

Bergson, and his consistent advocate, Deleuze, have led us down this course, where the cosmos is composed of images in movement (movement-images). Those advocating the centrality of phase conjugation see the brain and sensory apparatus[57] as transducers generating a distinct signature (reference) wave that can access spectra in constructing the three-dimensional spacetime within which we operate.

Ultimately, too, following the work of Bekenstein, 't Hooft, Susskind, and Bousso, it is the enclosed boundary surface of ecological niches that encodes, in two-dimensional form, all information residing within the bulk. Could it be that due to ongoing quantum fluctuations, some of this information sets loose in the form of a soliton, perhaps what Deleuze and Guattari referred to as a "line of flight," that traverses and reverberates in the field? To this we now turn.

---

55. See Jones and Mewhort (2007). They offer a "memory model" of a holographic lexicon. Their empirical study indicates that "word meaning and order information are stored together as a single pattern of elements in a distributed holographic lexicon." Word meaning, for these authors is a function of not only context, but by temporal positioning in relation to other words (ibid., 2). This is in accord with Lacan who additionally shows how the value of a signifier is only determined retrocausally ("retrograde"). See also Cox *et al.* (2011) who describe a holographic semantic structure.

56. Recall, from the previous chapter, Bradley suggests these are stored as Fourier descriptors within the brain.

57. Bradley (2006, 2007, 2009; see also McCraty *et al.*, 2002, 2004a, 2004b, 2009) has also suggested that the heart is a major emitter of frequency waves that encode information.

Political Economy, Schema QD, and "Point of Subjectification"

Of course, any discussion of consciousness must include its relation to political economy. Accordingly, let us provide some thoughts on this relationship. We return to our discussion in chapter 3 concerning Deleuze and Guattari's assemblage theory (1987; see also Patton, 2000: 94–108) and we will also integrate their notion that political economy can be conceived of as an axiomatic system.

Schema QD can be seen as accommodative to the notion of an assemblage. Assemblages are patterns, with degrees of stability, carved out of a plane of immanence. Recall the two orthogonal axes. The two poles of the vertical axis are identified as territorialization—all forces that contribute to the stability of a particular agent's matrix, distinguished by a distinct signature wave; and deterritorialization—anything that undermines this stability. The two poles of the horizontal axis include expression or instantiations of the quantum cloud of possibilities, and content, or bodies in relation to each other in a political economy. The specific matrix unique to a person can be conceptualized as her/his distinct diagram. With this, we have postulated that a distinct quantum signature wave is associated with each instantiated matrix of schema QD. As an assemblage, it is situated within a plurality of often overlapping assemblages at different scales (e.g., person-to-person, network, organizational, societal). Each assemblage can be mapped as a diagram, a relation of components (i.e., matrix). Each, too, has an inherent logic (function) that plays itself out as an "abstract machine." In our view, these generated lines of flight are affordances encoded with holographic information. When a more dominant abstract machine has taken hold we can conceptualize it as a form of a "pilot wave" (Bohm, 1983) and with the complicity of agents, a social wave function that traverses the field (plane of immanence). We can also conceptualize a more over-encompassing macro-level assemblage that we call a particular socius. This is distinguished by a boundary surface on which is encoded in 2-dimensional form, all information included in the bulk (i.e., spacetime happenings). Following Deleuze and Guattari (1987), capitalist economy is an axiomatic system, generating information, reflecting the inner logic of capital logic (see also Patton, 2000: 48, 94–103). An axiomatic system provides the various categories out of which everyday flow of energy is given form and meaning (i.e., territorialization). A political economy such as one that is driven by capital logic "captures" the distinct energy flows, or flux, within the plane of immanence. Undisciplined "flows" are brought within the centripetal powers of the axiomatic system (reterritorialization). Thus a "state-form," such as one built on capital logic in its state-regulated form, is a "machine of capture" (Patton, ibid., 99).

Theorems follow given axioms in syllogistic form. Truths are constructed syllogistically from the premises residing in the axiomatic system. This is an abstract machine suggestive of the correct construction of "reality." This suggests an underlying homeostatic driven system. We can also envision that the boundary surface of encoded information is the source of affordances, with the form of solitons reverberating within the bulk. Foucault's notion of the "disciplinary mechanisms" addressed three components of the disciplines (hierarchical surveillance, examination, normalization), to which we could add a fourth, inspired by Deleuze and Guattari (1987), an axiomatic system generating affordances, or waves encoding holographic information. By "direct perception" (Gibson, 1979) via Bergson's inattentive recognition form, affordances are instantiated and reified in everyday social practices. The continued instantiations of these affordances reifies dominant modes of understanding. Errant, or "mutant flows," are brought within the orbit of axiomatic systems (territorialization). Expressive forms could be first brought within a taxonomy of signs (linguistics) which reflects patterns or centers of vibratory movement in time and space, as Deleuze (1986, 1989) has shown in his work in cinema theory. This is then captured in a language (semiotics). This language, then, is populated by "order words" (Deleuze and Guattari, 1987: 79–81)[58] or in Lacan's view, "master signifiers" that call out a certain response; they speak the subject who is their user. Languages can become stratified, what Deleuze and Guattari call "regimes of signs." Within each, we can identify a particular structure by which signifiers are embodied.[59] A signifier, that according to Deleuze and Guattari (1987) normally remains in "continuous variation," or, in quantum symbols, $|S_i >$, tends to collapse to a distinct narrow range of expressive signifieds because of the forces of capture and axiomatization. These instantiated signifiers now coupled with a circumscribed range of possible signifieds, in turn, effectuate "incorporeal transformations" (Deleuze and Guattari, ibid.). That is, they can produce an immediate effect, such as in Gibson's direct perception, and in Bergson's inattentive recognition form. Literature in "speech act theory" (Searles, 1970; Austin, 1975), and in "language games" which are

58. "The only possible definition of language is the set of order-words, implicit presuppositions, or speech acts current in a language at a given time" (Deleuze and Guattari, 1987: 79).

59. Lacan (1991) has offered the idea of the four discourses (master, university, hysteric, and analyst), each of which produce a distinct flow of energy culminating in an embodied form of understanding ("truths").

said to limit what can be said in a language (Wittgenstein, 2009), support this view of language having distinct incorporeal transformative effects. It is phase conjugation dynamics by which resonance is created, an informational channel, where otherwise vibrating entities take on distinct expressive form ("reality").[60]

Deleuze and Guattari's (1987: 129–139) notion of a "point of subjectification" is also implicated.[61] This "point," the origination of a process by which a subject ("mental reality") is constituted as a speaking subject as an "I," is situated within an axiomatic system. An "I" is always situated in one or more overlapping assemblages at different scales. Bounded regions can also have nuanced axiomatic systems. That is, we conceptualize the instantiation of a particular QD matrix of an agent as the basis of the emergence of an "I," a speaking subject that takes up a position within these syllogisms in creating reality and him/herself. Recall, schema QD suggests a complex of interacting dyads: ego—ego-ideal, ego—Other, ego—community generalized other, ego-ideal—community generalized other, ego-ideal—Other, all situated within a background of a more universal, macro-level abstract generalized other (see figure 2). It is from this dynamic that a speaking subject, an "I," is instantiated that can take up residence in a particular discourse (regime of signs) to speak. Resonance (being in tune with the other) establishes information channels. In social interactions it is the exchange of the interlocutors' signature wave that produces an inter-

60. For another essay, we could reconceptualize Rossi-Landi's (1977, 1983) analysis of a homology between the commodity and linguistic form. He follows Pashukani's (2002) earlier explication of how the commodity and legal form have a similarity in development (homology). Rossi-Landi demonstrates how a signifier initially embodied with the desires of an agent as use-values, are transformed by a fetishism of the linguistic form into exchange-values taking on universal meaning (Milovanovic, 2003: 201–202). We would, following Marcer and Schempp (1998, 1999), show how the use-values "map" to exchange-values and are accessed as so via phase conjugation. Resonance, in short, is established between the two, with the uniqueness of embodied signifiers (use values) giving way to universal, abstract understandings (exchange values), devoid of the idiosyncratic embodiments of its users. Note, too, that this is a one-way process; the reverse "map from" possibility of phase conjugation is much more difficult due to ideological constraints that favor exchange-value.

61. Deleuze and Guattari (1987: 129) have said "various forms of education or 'normalization' imposed upon an individual consist in making him or her change points of subjectification, always moving toward a higher nobler one in closer conformity with the supposed ideal." They also see parallels with Althusser's classic statement on "interpellation" (ibid.,130), a form of "hailing," by which identity is confirmed onto those hailed. Here, too, we see Foucault's (1977) work on the disciplinary mechanisms, where his notion of normalization is one of the three components shaping the construction of social reality.

ference pattern, a standing wave, which is a communication channel for the construction of information.

Following Wendt (2006), a "social wave function" is a dominant information bearing wave reverberating within a bounded socius acting much like Bohm's "pilot wave." We follow his work in identifying, too, on the one hand, active monads that access this social wave function in creating (via phase conjugation) understanding and reality. Where a particular person's signature wave function resonates with the social wave function we have reification of the dominant order and stasis. Where there is discord (deterritorialization), we see the basis of alternative constructions. The state, functioning as an abstract machine, insists on reigning in these "mutant flows" (Patton, 2000: 98) or flux. In short, assuring "capture" of the flow inherent in the plane of immanence such as in wage labor. Lacan (1991), too, has identified the "discourse of the master" and in more hidden form, the "discourse of the university," which provide master signifiers from which a circumscribed reality is constructed. These "order words" produce incorporeal transformations in their use. The latter dynamic provides the basis for rethinking (deterritorializing) given practices of justice along the lines of a (reterritorializing) transformative justice, which we will take up in our final chapter. There we will emphasize the transformative potentials inhering in the person.

Conclusion

This chapter has been concerned with responding to the paucity, the vacuity, of any bona fide notion of agency in the criminology literature. It also responds to Jean Paul-Sartre's (1972: 40) critique of Bergson: "How do virtually represented images suddenly manage to encompass an 'I' by becoming 'present'?" It responds to his question, again critiquing Bergson, "how does an image turn itself into a memory-image?" Our position neither advocates a "transcendental subject," an agent that stands separate from a milieu within which it engages, nor a position that negates the effects of a nuanced, efficacious agent. Our schema QD, which is a quantized, de-oedipalized adaptation of Lacan's schema R, locates the dynamically interacting components of the agent in the framework of quantum holography. Much is yet to be done along these lines. We have also drawn from Deleuze, Guattari and Bergson in showing that within their frameworks, a reconceptualization can make room for a viable agent. We have shown that the speaking subject, "I," is a result of a complex process of phase conjugation and wave collapses. We have also suggested that molar assemblages capture the more dynamic flow of being human in circumscribed sets of signifiers. This is the principle of axiomatization to which Deleuze and Guattari (1987) introduce us. It is the "point of subjectification" from which a speaking subject is shaped, consistent with the normative system and its disciplinary mechanisms at work in a particular socius. We suggested that some signature waves reflecting the possible matrices of the agent tend to solidify toward more molar assemblage forms (more point attractors), some, more molecular (torus and strange attractors) in form. In the next chapter we want to move to applications of quantum holography principles to criminology and law. In our final chapter, we want to point out the significance of quantum holography in developing a transformative justice.

Chapter 5

Quantum Holography at Work: Applications in Criminology and Law

Introduction

In this chapter we would like to provide some applications to two areas: criminology and law. In our final chapter we will visit transformative justice. We do not set out to provide full-fledged theories, but rather seek to initiate an alternative framework for consideration within which these issues can be theorized. Accordingly, the reader will find possible directions for further analysis with alternative conceptual tools. The Rubicon needs to be crossed; we have only our static ontologies to lose. Our process-informational paradigm offers us the coordinates for this undertaking. We continue to raise objections to the contemporary entrenched theorizing that has embraced flat empirical and positivistic methodology. On the other hand, we do see some current research quite compatible with the ontology we have been developing. What we seek is to quantize it as well as integrate a holographic component. Let us first turn to criminology.

Criminology

The field of conventional criminology has been in crisis for some time due considerably to its insistence on a Newtonian ontology. It is time to break out. Arguably, cultural criminology with further integration of constitutive theory currently offers the best starting point in integrating quantum holography.

Cultural criminology finds its best-articulated framework in two comprehensive works: Ferrell *et al.*, *Cultural Criminology* (2008), and J. Young's *The Criminological Imagination* (2011).[1] Their key statements concern: how chang-

1. Jeff Ferrell's work has been most significant. See his early "Cultural Criminology"

ing political economy places some segments of the population at greater risk of criminalization; the importance of agency; practices of "othering"; the intermediary importance of subcultures; the allure of edgework and emotionality; and the emergence of a cyber-informational-cinematic/video consumer-driven society. We have no differences with their central themes, but offer complimentary explanatory analysis informed by quantum holography. The latter provides a physicality to much of the theorized concepts offered by cultural criminology. We also extend their central argument by incorporation of the works by Deleuze and Guattari (1983, 1987), Deleuze's two-volume books on cinema (1986, 1989), and constitutive themes (Henry and Milovanovic, 1996, 1999). Here we will focus on key themes of cultural criminology and how a quantized, holographic incorporation offers insights for further development.

1. Political Economy

Political economy, according to Young (2011: 92–103) has undergone major upheavals since the "golden age" of post WWII years that was characterized by "high employment, stable family structures, and consensual values" secured further by welfare state provisions, to new developments in the last third of the century and extending into the twenty first. The latter is characterized more by "structural unemployment, economic precariousness, a systematic cutting of welfare provisions and the growing instability of family life and interpersonal relations" (ibid., 92). In short, a society of increasing mobility, disruptions, greater competing values, risk, mass public consumerism, proliferating virtual realities, and inherent ontological and material insecurity (ibid., 92–93). "Late modernity," he tells us, appears with two tensions: the tension between diminished ontological security and a hyper search for identity, and the tension between the stress on the excitement of the moment against the backdrop of the packaging of leisure and the rationalization of the work place (ibid., 93). It is within this emergence that media practices create a blurring between virtual representations and the real (Ferrell *et al.*, 2004, 2008).

(1988). Other works include: Ferrell *et al.*, *Cultural Criminology Unleashed* (2004), and Ferrell and Sanders, *Cultural Criminology* (1995). It is with great regret that as I write this section, I have been informed of the passing away of Jock Young, for most of us, a passing of an era extending back 40 years to the first publication of the pivotal co-authored work, *The New Criminology* (1973), a work that had been, along with Richard Quinney's *Social Reality of Crime* (1970), the key influence in my own move to critical criminology while in grad school.

Let us quantize this change. We envision this change as a fundamental qualitative change in the socius, structured along the lines of assemblages. Forces of deterritorialization have gained dominance over forces of territorialization. Relations of bodies to bodies ("content") have undergone disruption in routine. Reterritorialization has been significantly enhanced by the mediated effects of the proliferation of images that are taking on a life of their own. The "expressive" dimension of the assemblage is calling forth new forms of narratives that seek to re-establish the ontological security of the better days of the "golden age." The plane of immanence, a plane of energy in movement, is being restructured. The emerging form of capital logic captures this movement within new expressive categories, while subjects attempt to break out of its constraints in such forms as edgework, one avenue being crime, while simultaneously seeking more stable identities such as has been established in the literature on "identity politics."

The plane of immanence is being re-axiomatized. This axiomatic system becomes the basis of affordances reflecting new expressive forms of capital logic in late modernity. This is the function, or logic of an emerging societal-wide assemblage, an abstract machine. Take the 14th Amendment to the U.S. Constitution as an example. It includes two clauses. One is known as the "equal protection clause," which, paraphrasing, states equally situated should be equally treated," the basis of formal equality. This is part of an axiomatic system. Theorems derive from this central postulate. Everyday reality construction invokes bits and pieces of the axiomatic in constructing meaning in everyday interactions. This in turn reinforces the axiom as a taken-for-granted truth. In other words, created affordances, as wave functions, or more specifically, holographic-based information bearing waves transverse the sociopolitical field. Wave functions, we are reminded, are clouds of possibilities. It is the collapse of the wave function in relation to the constitutive work found by agents diagramed by schema QD that produces a particular instantiation of what we see as "reality."

Media practices are active in molding virtual realities from this primordial information flow (Ferrell, *et al.*, 2008: 122–158; 2004). Media is active in constructing packages, reality bites that are disseminated instantly throughout the socius via the exponentially expanding technological machinery. Information creation entails "loops" and "spirals" whereby it is constantly massaged and packaged as virtual reality for a willing public consumption. Many forms of cop shows and television criminal justice dramas provide particular, one-sided ways of collapsing a wave function, suggesting a truth in virtual realities created. Identities are polarized—(good guy, bad guy), so too narratives (lawful, unlawful), cause (linear, nonlinear), guilt (yes, no), emotions (rational, irrational)—from which emerge contrived truths in virtual forms, competing against the actual. What is real and what is virtual run after each other, loop and spiral, continuously driven by mass consumerism. The newly emerging socius, or macro-level assemblage, is therefore, an axiomatic system, expressed by its abstract machine, generating affordances, wave functions that encode holographic information. The receivers, or percipients, are everywhere assaulted in all aspects of their lives by this relentless logic, which, as Deleuze and Guattari (1987) have demonstrated can also be seen as a "point of subjectification," a point from which waves emanate, traversing the social field, and which become the basis of identity construction and everyday expressions of an "I" in discourse. Everyday reality-constructing practices entail active constitutive work by which postulates from the axiomatic system enter the construction process of perception-images. These loop and feedback, to constitute, and in its continuous practice, reify, what are perceived as given truths in the political economy. It is phase conjugation that accounts for how these incoming waves are negotiated by our sensory apparatus, particularly our brains, transducers, which both decode and remit information bearing waves by way of the organizational logic of schema QD.

Although this movement clearly expresses a deterritorialization and a reterritorialization, a "capture" by the axiomatic system, we also see the positive possibilities. "Lines of flight" that are ever deconstructing the order of the past, are also the potentials for new assemblages. That is, from the ongoing flow of the plane of immanence alternative blocs of space time can be cut reflecting, for example, the many creative possibilities inhering in the wave function, a cloud of possibilities. Bergson's "creative evolution" suggests that disruptions and attentive recognition are moments for alternative constructions. Deleuze (1986) shows how the latter is the basis of the "power of the false," the ability to create, not discover truth. Truth is only discovered in an accepted axiomatic system. The power of the false is the ever-present ability to construct perception-images differently.

2. Question of Agency in Cultural Criminology

This brings us to the question of agency. Cultural criminology brings back the importance of agency without, however, providing the wherewithal of an inter- and intra-subjectively constituted agent. A long endnote in Ferrell *et al.* (2008: n.1, pp. 22–23) perhaps provides the clearest expression of agency from this perspective. Here stressed is the potential creative nature of human action. It is, following Matza (1969), when agents are pacified that they follow a script and appear determined, what Bergson would attribute to passive recognition. Young (2011: 84), too, follows Marx's classic thought that agents are active creators of reality but often not of their own choosing. We return to our schema QD. It can offer cultural criminology a better conceptualization of the human being, but rooted in quantum and holographic principles. Phase conjugation and the collapse of the wave function can be incorporated in cultural criminology's explanation of the physicality in the process of instantiations of virtual realities. Perception-images, the basis of motor action, have determinants in the play of the various dyads developed in our earlier explication of schema QD, all being further played out in a sociocultural field permeated with traversing affordances that holographically encode information. Cultural criminology cannot find its basic ontological principles in Newtonian physics. Rather, it already is a process-informational model that is in need of being quantized. In our own constitutive approach (Henry and Milovanovic, 1996, 1999; Milovanovic and Henry, 2001) we show how agents often invest in energy to make a difference, to categorize and hierarchize. In other words, blocs of space time are being created which reduce the other's potentials. Young, of course, attributes this to a growing ontological insecurity in late modernity.

3. Othering

Inherent ontological insecurity of later modernity, says Young (2011: 63, 197; 2007), produces distinct forms of "othering." Conservative othering is a form of categorical stigmatizing; liberal othering, a more hidden but pervasive form that "diminishes" and "distances" the Other—the Other being seen as lacking in control, education, civility, rationality, etc. Our work (Milovanovic and Henry, 2001) has referred to this as harms of reduction and harms of repression, the diminishment and denial of agents in their abilities to make a difference. Agents who embrace this ethos are excessive investors in harm. Matza (1960) as well as more recently Young (2011) have suggested that this act of exclusion can engender humiliation, which has one possible avenue crime. Matza (1960) has called this an act of desperation. Young (2011: 109) has referred to this "as a breaking through of restraints, a realization of immediacy and a reassertion of identity and ontology."[2]

Previously (Chapter 4), we have argued that we could conceptualize three others which reflect intra- and inter-subjective constitutive elements: the face-to-face Other, a community generalized other, and a more universal abstract generalized other. Each, too, can be seen in terms of interactions amongst

2. Halleck (1967), applying a psychoanalytic perspective has also argued that crime can be seen as identity confirming activity.

dyads that are fundamentally about emitting and remitting wave frequencies while accessing a holofield encoding spectra, wave frequencies that can be activated in producing holographic images. A person's unique emitted signature wave activates and creates interference patterns with these incoming frequency waves emitted in the social field. The community generalized other and abstract generalized other can be seen, then, as more stable spectra encoded in the holofield. Where resonance takes place (phase conjugation) we have informational channels established. The "I," or speaking subject, is both an after-effect, and, due to backward referral of quantum information, a causative agent in the end construction. Young's "other" can be better reconceptualized as a derivative of a more pervasive, societal-wide abstract generalized other that has its determinants in late modernity. Perception-images created, as well as a speaking subject, are results of this process. This suggests, alternatively, that an ethics could be articulated based on an othering suggested by: Buber (1971), in the "I-thou" rather than an "I-it" relation; in Levinas' (1969) suggestion of the infinite duty of care for the other; in Nietzsche affirmative action (Deleuze, 1983); or in Spinoza's enhancing the capacity to effect and be affected (1985; Hardt, 1993). To this we will return in our final chapter on a transformative justice.

4. Relevance of Cultural Practices

Cultural criminology strives to "bring culture back in" (Young, 2011: 203). "Objective conditions," he argues (ibid., 77), "are interpreted through the specific subcultures of groups involved." Specifically, narrative constructions are key elements in subcultural constructions of "objective" conditions. Thus the turn is to how, on the one hand, reality and meaning is constructed, and how, on the other, it is distorted, eliminated, transformed. To this end, Young, and to a considerable degree other cultural criminologists, embrace Matza's (1969) statement on "naturalism," trying to be true to what one is studying. Nevertheless, this naturalistic stance assumes some objective reality already in existence (Gubrium and Holstein, 1997: 19, 38) and to understand it, in this stance, one has to "be there" in order to be appreciative of this "reality." Alternatively, and consistent with our quantum holographic approach, reality rather than being seen as discoverable could be seen as an ongoing co-construction by participants (Gubrium and Holstein, 1997: 38–56; Wieder, 1988; Henry and Milovanovic, 1996).[3] The focus is

3. Consider Gubrim and Holstein (1997: 38): "Reality is accomplished rather than merely experienced ... that there is no 'there' until it has been constructed."

more on how members create narratives by referencing unstated norms. They ongoingly create the normative order through their everyday narrative constructions that naturalists take for granted. In this approach, reality is co-construction. The everyday taken-for-granted constructions reify dominant orders.

Our schema QD indicated how vibratory energy states in a plane of immanence are accessed, decoded, and returned by a reference wave which includes the unique signature wave of the agent to form interference patterns and an informational channel. The "out there" is created at each instance. Vibratory energy is not in a fixed state, but rather, as a quantum wave, reflects potentiality for expression, a "cloud of possibilities," until collapse to a particular instantiation. The perception-image that results, and the speaking subject ("I") are emergent from this process. It is not linear, nor strictly determinable; rather, each logon created in ongoing interactions reflects indeterminacy, for one cannot at the same time capture spectra *and* spacetime. Thus, subcultural member's accounts/narratives actively constitute their reality, be it slightly nuanced on each occurrence (ibid., 44–45; see also Henry and Milovanovic, 1996; see also the work of ethnomethodologists, Garfinkel, 1981; and social constructionists, Berger and Luckmann, 1967). The apparently appearing circularity is accounted for by the backward time referral of quantum information. The "I," unlike positivist thought, is both affect and key determinant of the reality s/he creates. Ferrell *et al.* (2008: 137) begin this departure in their summary of rap and gangster rap music to which we will now briefly turn.

The question of the relation of street codes, gangster rap music and identity has been insightfully developed by Kubrin (2005). What is most significant about this study is that it is a holistic, constitutive approach integrating political economy, subcultures, and social identity. Kubrin traces the emergence of gangster rap to the early 1990s to artists from the West Coast (ibid., 360–361).[4] Gangster rap, unlike rap music generally, "is specifically interested in the black underclass in the ghetto" (ibid., 361).[5] Gangster rap music is also a form of storytelling. It is not, Kubrin continues, a "cause" of violence, but helps constitute violence oriented identities. The social context of a street code and gangster rap can be situated in the degradation of inner-city communities which has its roots in globalization, descending industrialization, a history of

4. For the west coast beginnings, see Keyes, 2002; Perkins, 1996.

5. See also Keyes, 2002; Rose, 1994; Perkins, 1996. Ferrell *et al* (2008: 71, italics in original; see also Young, 2011: 109), explaining acts against the law within deprived areas, are often "*experienced and perceived* in an affluent society as an *act of exclusion*—the ultimate humiliation in a society defined by wealth and consumption. It is an *intense* experience, not merely of material deprivation, but of felt injustice and personal insecurity."

discrimination, and a punishment-oriented criminal justice system (Kubrin, ibid., 362).[6] Out of this, a street code emerges, a basis of a subcultural social order where identity and respect become paramount virtues. At this point Kubrin varies from traditionally inspired research on culture/subcultures which embrace a naturalistic approach, an approach that argues that gangster rap lyrics emerge from a normative order found in the street code which then acts as "subcultural directives" (ibid., 365). Instead, Kubrin, follows the work of Gubrium and Holstein (1997), who in turn are heavily influenced by the classic constitutive thesis developed by Wieder (1988) in his examination of parolees in a halfway house. This constitutive approach has its roots in ethnomethodology and social constructionism.[7]

In this alternative view, Kubrin argues that reality does not exist independently, objectively as "out there," but is constituted in everyday interactions and narratives that attempt to make sense and provide a degree of predictable order by members of that subculture. In this view, the street code is constituted in everyday narrative constructions. This everyday narrative construction, then, constitutes the normative order. The "telling of the story" is equivalent to "doing things with words" (Gubrium and Holstein, 1997: 49; Wieder, 1988). Members create borders by "talk and interaction" (ibid.). Thus, "both the street code and rap lyrics are constitutive elements of contemporary black urban culture ... [they] create the sense of a normative climate of violence. They provide sometimes graphically detailed instructions for how to interpret violent, degrading conduct and in so doing create possibilities for social identity in relation to violence" (Kubrin, 366). Street codes and rap music thus provide "interpretive resources" out of which everyday reality can be constructed in narratives.

Let us now turn to a quantized and holographic incorporation. First, we borrow from Deleuze and Guattari (1987) and argue that inner city communities can be conceptualized as assemblages that can be articulated by a diagram identifying the existing pattern of energy vectors and their etiological sources (e.g., territorialization). Deleuze and Guattari (ibid., 313–323) have suggested that any primordially vibratory milieu experiences a "periodic repetition" which manifests instances as blocs of spacetime producing a code of repetition, all of which can be recognized by a rhythm, resonance, the "refrain," a harmonic

6. See, for example, Rose, 1994; Wadsworth, 2004; Wilson, 1996; Keyes, 2002; Hawkins, 1985.

7. See also our work on constitutive criminology (Henry and Milovanovic, 1996, 1999) and its basis in ethnomethology, social constructionism, and Gidden's constitutive approach.

tune.[8] Territorialization is then about how these milieus evolve into territories and strata, more enduring social spaces. Out of these assemblages, expressive forms are constituted such as street codes. These milieus, or inner city communities, then, are also geographically bounded with relatively stabilized street codes. These can be envisioned as an axiomatic system which offers the basis of theorems and the elements out of which reality can be constituted. This is the logic of an abstract machine and its generated logic, and can be seen as the bases of affordances (Kubrin's "subcultural directives"). Affordances are spectra, or interference patterns encoded with holographic information. A more dominant form of expression in this assemblage that follows an axiomatic system with its syllogism is a street code; from this, following a syllogism, emerges rap and gangster rap music.[9]

To be more precise, we need to draw from both classic holography (Gabor, 1948) as well as quantum holography (Gabor, 1946; see also Bradley, 2010, 2011). Classic holography would indicate how information is encoded by way of interference patterns (spectra). These are stored non-locally in the holofield (ZPF). Within the boundaries of often segregated, underclass communities indigenous spectra are established reflecting reoccurring behaviors. As Bradley (2010: 129) notes, "as behavioral commonalities shared by all members, these particular modes of conduct establish a bandwidth of behavior and activity that create an energetic resonance or vibration—a 'vibe'—that uniquely characterizes the group."[10] This reflects "socio-bio-emotional commonalities" (ibid., 131).[11] A community's distinct signature wave would, following Bradley (2010: 126), be "enfolded as tacit information—a latent commonality—in the energy spectra of all interactions among group members, even in the smallest

8. Bradley (2010: 127) has noted that a "collective identity ... derives from the members' proprioceptory experience of the patterns of physical sensations, feelings, thoughts, behaviors, and interactions that give the collective its familiar characteristic shape, texture, and relational form—a tacit *configural* reality—as an ontological object of perception."

9. Seemingly, it might seem that this is in accord with Sutherland's (1974) "differential association theory," that an "excess of definitions" will unilaterally lead to violence. However, it is often forgotten that he tempers this with his idea that "excess definitions" vary in intensity, priority, duration, and frequency. Following the work of Delanda (2006), we would also add critical timing and differential threshold levels. The latter, then, suggests a more nonlinear flow of possible responses to an "excess of definitions" favorable to crime.

10. This could be conceptualized as "focal concerns" (Miller, 1958) or "excess of definitions" (Sutherland, 1974).

11. Bradley (2010; Bradley *et al*, 2010) has said "socio-bio-attunement establishes a tacit field of energy oscillations that radiate throughout the group, permeating all aspects of group-related behavior and connecting all members at a common energetic bandwidth."

interactional episodes." Within the boundaries of a group, a "configural reality" (ibid.) is established. This is equivalent to an unarticulated, background relevancy space, conceptual space, or possibility space (Delanda, 2006: 29–31; Garfinkel, 1991; Gardenfors, 2004). Drawing from Delanda and dynamic systems theory, this is a phase space distinguished by "degrees of freedom" or possibilities, replete with distinct attractor forms to which the system tends. It is a "socioaffective field" (Bradley, 2010: 138) which functions to "unify and activate affective attachments, arousing each member's emotions and thus their potential energy—their propensity for expending bio-emotional energy in action."[12] Disenfranchised, segregated communities create street codes, a "tune," or "refrain." An expressive form of this assemblage reflects commonalities in energy expenditure and associated frequency patterns consistent with the "essential character" of this field (p. 145).

The "constellation of plans and modes of expression" (p. 145) would include gangster rap music as a concrete form of expression arising from street codes which provide constitutive elements in its construction. Members of this collective emit their own unique signature waves in this field, frequency waves encoding holographic information. Street codes can be conceptualized as interference patterns, or spectral encoded information. Street code waves, representing the configural reality found in that milieu as "waveforms of oscillating energy" (143) are received by members, decoded via the Fourier transforms, recoded with the inclusion of the unique signature wave of the member, and re-emitted outward creating interference patterns some of which follow the same pathway as the incoming waveform representing street codes. Between the incoming and outgoing wave forms, we envision the dynamics of schema QD being mobilized. That is, Bergson's zone of indeterminacy is implicated. His

---

12. He (ibid., 139–140) also notes that negative emotions such as hate, fear, and anger—not in short supply in responses in segregated and disadvantaged communities (see note 5 above)—tend to undermine potential energy conversion and thus lead to stasis and ossification, in our conceptualization, molar forms of assemblages, or enduring subcultures with higher prevalence of conceptual (possibility) spaces suggesting and supporting violent resolutions. Street codes, then, are expressive forms of this conceptual space. On the other hand, Bradley tells us, "mutual admiration, awe, or love create a kind of harmonic resonance" and thus, in our view, a molecular form of assemblage, suggesting moments of transformation. Bradley suggests if the more bountiful form of energy is not sufficiently constrained by some control apparatus, chaos will emerge producing destabilization in the community. Some balance, in short between flux and control is necessary. However, we could also consider non-linear dynamics and the possibilities of social organizations based on jazz or radical jazz music (Holland, 2011) as a model which would lead to some suggestions of a more humanistic society, a topic to which we will return in our last chapter on transformative justice.

"circuit diagram" points out how memory images, stored "out there" (holofield) are accessed and then checked against a series of recollection-images for a fit. We add here that the various intersecting dyads found (in matrix form) in schema QD shape this process of finding an appropriate match.[13] Momentarily we would have quantum coherence as many possibilities sit side-by-side until the collapse of the wave function produces one perception-image, which as a waveform of oscillating energy incorporating the unique signature wave of the member (percipient), is re-emitted outward as a virtual image. With the subsequent interference pattern created with the incoming wave, an information channel is opened, a 3-dimensional object is created. In this model constitutive effects entail backward time referral of quantum information whereby the "I" that becomes the interlocutor is both the author and result of the final perception-image created. That is, as information is sent backward in time, an "I" is taking form which then takes up residence in narrative constructions as the speaking subject. Members of disenfranchised segregated communities intermittently make use of the street codes in constructing their perception-images and by so doing reconstitute and reify the very structure to which they refer. As Bradley (ibid., 147) tells it, "by attuning all members to the same resonant socio-bio-emotional frequency, the group generates a powerful collective receptive field through which energetically encoded information is accessed and amplified due to a stronger resonant feedback loop, both to the field of the group and to that of the individual member." It is within this dynamic that accessing street codes provides the grounds for a fertile basis of lyrics for rap and gangster rap music.

Unlike Bradley, however, who argues that the collective signature wave of some group becomes encoded in the brain of members of a group as Fourier descriptors, we embrace a more Bergson-Laszlo position that it is located (encoded) in the holofield as relatively stabilized frequency patterns (holograms). These are accessed by a person's unique signature wave that is shaped by the dynamics of schema QD. Group members have similar abilities to access this nonlocal energy spectra (holograms) reflecting their configural reality. Thus in this view, the brain is a transducer, receiving frequency waves, and emitting its own signature wave accessing the holofield, and returning a virtual wave a composite of the signature wave and incoming frequency wave, back to the source, producing an interference pattern, a phase conjugation. This is an informational channel. This instantiates (constitutes) the particular emitting ob-

13. Recall, a schema QD matrix undergoes a collapse of the wave function producing a distinct signature (reference) wave that accesses the ZPF. An instantiated "I" in narrative constructions is both a result and retrocausal effect.

ject as a "reality" we consciously perceive "out there." But this does not yet complete the picture. We turn to quantum holography and the matching of the two forms of holography (classic, quantum) to explain harmonic resonance, or a felt sense of commonality in the group, and how rap and gangster rap music, one expressive form of the street code, resonates or produces a common "vibe" with community members. Suggested is that the linearity of classical holography and so too classical constitutive theory overlooks momentary constructions that are spacetime constrained; that is, the logons constructed are but possible versions of the embedded codes within the dynamically evolving milieu.

Turning to quantum holography, it would explain how everyday interactions within the disenfranchised, segregated community are momentarily encoded as logons, or "time-constrained" quantum bits of information.[14] In our application, we theorize that these logons are bits of information from which narratives and reality are constructed. Gabor (1946), we earlier explained, outlined the basis of the wherewithal of logons in his investigation of message transmission to determine the most efficient message (minimal uncertainty of intelligible information) in the least space. Logons are momentary and ongoing perception-images that are created in member's daily activities. Members of marginalized and disenfranchised communities as others in various other communities create logons in their everyday interactions.[15] Whereas Gabor's 1948 article was more about image processing, and for Bradley, the basis of the construction of more static spectra (classic holography), his 1946 article spells out a more dynamic process of holographic construction accounting for every momentary constructions of information. Logons are "spacetime constrained holograms" (Gabor, 1946; Bradley, 2010: 148). Bradley (1998; 2010) offered

14. As Gubrium and Holstein (1997: 49), referring to the classic study by Wielder, 1988) of convict codes, state, "the code was not a separate and distinct normative structure, available to the researcher for explaining why residents acted as they did. Rather, it was variably conjured up by the residents themselves, who used it to account for matters in need of explanation."

15. There is variability, or differential access in this process. As Gubrium and Holstein (1997: 50), referring to Wierder's (1988) study of inmates at a halfway house state, "when the code was not in use, neither were its particular normative order and resulting social categories. In talk and interaction, the everyday reality and social effects of deviance were not constant artifacts of an ever-present set of norms. Rather, the normative order (in this case, the convict code) was an indigenous resource for invoking normative accounts for whatever residents did or did not do in the halfway house. In other words, applicable social categories and their normative order were phenomenally present, in practice, only upon the occasion of their use."

the idea of "flux," a flow of energy, what Deleuze and Guattari (1987) locate on a plane of immanence. This is shaped by "control," which we previously saw was hierarchical power. These are two perpendicular axes (orthogonal). They spell out a phase space within which logons are blocs of spacetime. We, however, revise the notion of "control" by substituting the notion of constitutive construction; members recreate systems of control in their everyday usage of key components of the axiomatic established in that field. They constitute the very control about which they speak in conversation, in narrative form.

Thus logons reflect the intersection of relatively stable spectra characterizing the street code encoded within the holofield ("socioaffective field") with bio-emotional expenditure of energy in movement. These logons are the basis of the construction of gangster rap music. They become a part of the perception-image generated. An interference pattern, or harmonic resonance is created, a "vibe," between lyrics of gangster rap music and the configural space assumed by street codes. Thus ongoing pattern matching (Bradley, 2010: 148–149) between the relatively stabilized spectra encoded within the holofield of a community and the moment by moment creation of logons, the basis of subsequent work in the production of street codes and gangster rap music, produces a harmonic resonance, phase conjugation, amongst members within this cultural milieu. This extends, nonlocally to other less disadvantaged communities, particularly with the aid of media's insatiable desire for producing ever new sellable products, and to those who are the recipients of harms of reduction or repression.

Returning to Kubrin (2005: 361), he argues that his approach does not suggest that gangster rap directly causes violence. Rather, examined (ibid.) are "the more subtle discursive processes through which rap helps to organize and construct violent social identity and account for violent behavior." Structural factors of inner city disenfranchised communities are the basis of subcultural adaptations within which is encoded a reflective distinct street code offering an "interpretive environment" (ibid.). This has been referred to as "templates for action" (Schwartz *et al.*, 2004), or readily available responses to particular stimuli. Gangster rap lyrics selectively encode street codes, some of which justify violence. Kubrin argues that the code of the street "resonates" with gangster rap music (ibid.). But this is not unilateral; there exists a "complex and reflective relationship" amongst the three terms: street codes, gangster rap music, and social identity.

We have already examined the constitutive nature of this relationship. Here we want to extend the notion of "causation" to include relevant concepts from schema QD, Deleuze and Guattari, and quantum holography. Focusing on schema QD, we can now see that gangster rap music provides the linguistic coordinate system within which an "I" can be situated, from which the subject can speak. Members find distinct understandings of the Other within street codes as well as the lyrics of gangster rap music. The ego-ideal, I(0), how one views and feels oneself as a likeable person, revolves around receiving or not of respect; an extreme and definitive corrective measure, a template for action, is violence. The memory images of a member's symbolic and imaginary order, encoded within the holofield, provide the elements out of which recollection images are created. Together, this is a relevancy space, or conceptual space within which everyday reality is constructed. Affordances, or circumscribed action patterns (i.e., templates for action),[16] that circulate within this domain are frequency patterns encoding holographic information. Through phase conjugation, resonance is established between the lyrics of gangster rap music, street codes, and social identity. The collapse of the superpositioned possibilities (recollection images) produces one instantiation of an otherwise vibratory entity with potentialities of expression, into one virtual image that is returned to the object, constituting it in its distinct form (reality), e.g., an instance in which the construction is one of a different gang member dissing me, and grounds for response. Gibson's "direct perception" could account for passive recognition and more direct action ("templates for action"); whereas active recognition could produce a pause within which other action patterns are momentarily entertained. We have already argued that "interrupters" employed by "Cure Violence" in Chicago and other cities work on actualizing a different interpretation of the what happened and thus collapse the wave function into a different, often, nonviolent expressive form. But we also hasten to repeat, that this is a constitutive process: a person's "I" emerges from its constructions of the situation that borrows from the highlighted relevancy/conceptual space found in street codes and gangster rap music.

Deleuze and Guattari's work on causation as it relates to assemblages is a central ingredient in understanding how street codes, rap music, and social identity are connected. Delanda (2006), from Chapter 3, has provided some poignant commentary on their notion of causation. Rather than linear causation, which assumes distinct objects in relation to each other, moving in a continuous time, Delanda first argues that the internal structure of an entity plays at important role as to how it will respond to some perturbation or stimuli. This would be the "intra-subjective" components of the inter- and intra-subjectively constituted subject. Delanda states that each entity operates internally with different thresholds at which it will be affected and to which it will react (ibid., 29). He offers, then, a nonlinear notion of cause, "catalysis," which draws from

16. Sometimes it is referred to in the literature as sensorimotor schemas, suggesting a stabilized, patterned connection between a sense input and a motor response.

Deleuze's notion of "quasi-causality."[17] Similar stimuli, then, can have different effects since there are different thresholds. An affordance, as an invariant structure (Gibson, 1979), is a wave function, a cloud of possibilities, and still needs to be "collapsed" to one perception-image. Not all members of a disenfranchised ghetto will react to gangster rap music in an identical manner; some will, some will only in part, some will not at all. This is not a reliance on a form of "statistical causality" which focuses more on statements about populations, but rather, each entity has its own powers to be affected and to effect others. For Delanda's understanding of Deleuze and Guattari, assemblages can be identified with "catalytic causation," since, like persons, each has the capacity to effect and be affected partly due to its internal organization and threshold value. Earlier we saw how in the Penrose-Hameroff model of "objective reduction" what was posited is the idea that increased intensity would eventually lead to a collapse of the wave function, objectively, producing one rendition of reality. Here, rather, we argue building on Delanda that the objective reduction always has subjective components. Delanda (2002: 85–116), building on the work of Arthur Winfree (1987), also notes the importance of two additional factors, beyond intensity, in the quasi-causal model: critical timing, and duration.[18] There are differentials in the appearance of these factors that must be considered along with the internal organizational structure of the person or assemblage. This comports with our model of schema QD and with Deleuze, Guattari, and Delanda's notion of non-linear causation.

Kubrin's theses that structure, street code, gangster rap, and violent identities are intertwined in complex and constitutive ways, could be augmented with quantum holographic principles. Gangster rap music, as a form of a "point of subjectification" (Deleuze and Guattari, 1987) can, with more, produce a speaking subject, an "I," a distinct social identity that is instantiated in time and space, immersed in a particular background relevancy space (lyrics of gangster rap music). Phase conjugation takes place between street codes and gangster rap music, some of which, in direct perception in conjunction with the determinants of thresholds, critical timing, duration, priority, frequency, and intensity increase the potentials for violence. Where ongoing matches in frequency waves occur amongst members of a disenfranchised urban community we have harmonic resonance felt as a basis of solidarity or a commonality in expectations, understandings—a normative "ways of doing things."

17. See also Delanda's (2002: 75–81) more detailed explication of quasi-causality.

18. For another day is the question and response to what degree is Winfree's notion of "entrainment" compatible with quantum entanglement.

With Young (2011: 85), structure does not linearly construct gangster rap music and violent identities. Gangster rap, as a form of storytelling, draws its material from street codes that describe a relevancy background space from which elements are pulled in creating lyrics. The normative order, therefore, is constitutively constructed and continuously reified. Constructing lyrics ongoingly reconstitute street codes as seemingly self-evident normative structure, which then provides the interpretive structure for yet new lyrics. For Young, "it can be as much an act of acquiescence as of rebellion." We have seen that as an affordance, reflecting an axiomatic system, direct perception, and passive attention, could indeed collapse the wave function in dominant subcultural forms producing a seemingly appearing linear connection between structure, street codes, gangster rap, and violent identities. We have also shown, by referencing the work of "interrupters," that an active recognition can be mobilized (Bergson's "zone of indeterminacy"), which potentially could lead to an alternative collapse of the wave function. With Young (ibid., 104), we argue that rebellious acts are not necessarily meaningless. Street codes and gangster rap, following Young (2011) and Ferrel *et al.*'s (2008) analysis of subcultures, are attempts to respond to the ongoing discrimination and system of harms of reduction and repression. Viewing the members of disenfranchised communities through the lens of rational choice theory would have it that members are misguided in the choices they create, are equivalent to miscalculating financial advisors in the utilitarian thought process they pursue (Young, 2011). To which Young's witty response is: "the only sort of criminal who acts like a financial advisor is, in all probability, a financial advisor." Thus rational choice theory overlooks "foreground factors," emotionality and adrenaline rushes that can be engendered and cultivated as a form of response to structurally induced forces of exclusion, humiliation, and disrespect (ibid., 108; Ferrell *et al.*, 2008: 51; Katz, 1988, 1999). To this we now turn.

5. Allure of Edgework and Emotionality

Cultural criminology seeks to resurrect emotionality as a key force in agency and deviation from the normative order. Late modernity, apart from its disillusionments, unkept promises, and disintegrative effects, continues to place in the forefront rationalism as a central value of progress and change. Rational choice theory in criminology is but one manifestation. All are seen as calculating machines exercising or having the capacity to exercise rational choice. What is left out of the formula is emotionality. Cultural criminology makes the claim that it is on par with other faculties in actions by agents. Several key works have set the theoretical agenda: Stephen Lyng's (1990, 2005) classic statement on edgework; Jack Katz's (1988) work on the seductive nature of much crime, and his subsequent work (1999) on emotionality in action; and Patrick O'Malley and Stephen Mugford's (1994) and O'Malley's (2010) of edgework integrated with political economy. Much in the literature already implies the importance of excitement: Albert Cohen's (1955) study of expressive rather than instrumental behavior of delinquent boys; Seymour Halleck's (1967) study of excitement in the planning and execution of crime; Walter Miller's (1958) study of the "focal concerns" of lower class cultures, elevating adventure and excitement to a high premium. More recent work has and is appearing on a variety of edgework experiences (Lyng, 2005) including terrorism (Cottee and Hayward, 2011). Some have raised the issue of gender bias. Several more gender inclusive studies have appeared in the literature which have shown the application of edgework more generally (Lois, 2005; Rajah, 2007; Gailey, 2007; see also Ferrell *et al.*, 2008: 72–74). There have also been key statements in the sociological and philosophical literature on the importance of emotions (Sartre, 1972; Hochschild, 1983; Scheff and Retzinger, 2002; Denzin, 2007; see also Gubrium and Holstein's, 1997: Chapter 4). All this material places emotionality and the visceral in the foreground, not in the background. The latter focusing on more macro-level forces said to be at play in behavior, including crime (see Ferrell *et al.*, 2008: 56–74).

What is still needed is a quantized adaptation of emotionality. Our schema QD included the affection dimension. Deleuze, who was inspired by Bergson in this area, did recognize that emotionality can be situated between perception-images and action-images. Affections reside in the body, whereas perception-images are encoded "out there," in the holofield. But neither Bergson nor Deleuze sufficiently integrate inter- and intra-subjective components. Norman Denzin, in *On Understanding Emotions* (2007), and Katz, in *How the Emotions Work* (1999), are best situated for a quantized reconceptualization. According to Denzin (ibid., 3), "a person cannot experience an emotion without the implicit or imagined presence of others" and "emotional experiences are social experiences and require the presence of others (real or imagined) for their occurrence" (ibid., 92). Accordingly, emotionality is best understood in processes of interaction. Further, "a society and its laws [and normative order] enter directly into the organization and experiencing of everyday phenomenological life through the structuring of the moods, emotions, and emotional experiences felt by its members" (ibid., 24). Jack Katz (1999), independently from citing Denzin's work, sees emotions as a fundamental resource residing within the body which can be exploited in everyday interactions. Emotions, he tells us (ibid., 7) are "embodied foundations of ourselves." Sometimes we conjure them up to be strategically used in social interactions, sometimes they take over our very being.

This is a critical turn in the literature, particularly in criminology. But we also turn to quantum holography to provide the physicality of the issue before us.

Bradley, once again, has done groundbreaking work. He (2011) conceptualizes a "bio-emotional field," "bio-emotional energy," "socio-emotional experiences," "affective attachments," "affective energy," and a "socioaffective field." In short, the socioaffective field representing "flux"[19] or "potential bio-emotional energy," which, along with "control" determine a phase space within which attractor states flourish, or areas within which a particular energy expenditure will take on the form of a logon, a quantum unit of information. Following Laszlo (2007), however, this field is a holofield. Bio-emotional energy can be seen as vibrational and radiates throughout a particular bounded region. It is by being in tune with this energy that the body produces particular affection-images. Denzin (2007: 50), in the same direction, tells us, "emotions are not things, they are processes." In other words, emotional energy reverberates throughout the field. It is via phase conjugation that this energy form finds harmonic resonance with bodies. Thus, returning to the work of Kubrin (2005), and we shall show, progressive edgework inspired work on terrorists (Cottee and Hayward, 2011), what produces resonance with an established street code, or terrorist code, is, following Bradley (2011: 216), by way of "attuning all members to the same resonant socio-bio-emotional frequency [and] the harmonious group order that emerges from this bonding pattern generates a self-reinforcing collective field of coherent bio-emotional energy which amplifies the group's identity signature [unique signature wave embodying holographic information] ..." Heart rhythms play a central role, according to Bradley (2007, 2009, 2010: 136). The heart, in much literature, is equated with emotions. Bradley's review of the empirical literature, indeed found that the heart is the most powerful emitter of frequency waves (electromagnetic radiation).[20] Empirical evidence (McCraty, 2002, 2004; McCraty *et al.*, 2004a, 2004b, 2009; Radin, 2004, 2013) indicates it records and reacts to incoming frequency waves prior to the brain and consciousness recognizing them.[21]

Let us take as an example an insightful recent article by Simon Cottee and Keith Hayward (2011), "Terrorist (E)motives: The Existential Attractions of Terrorism." "Terrorism," in their view, "is as much a site of individual self-drama and self-reinvention as a tactical instrument for pursuing the political goals of small groups" (ibid., 963). Terrorist cells are distinguished by social bound-

19. Flux is accommodative to Deleuze and Guattari's "plane of immanence."

20. Bradley (2007: 83) states "it is likely that the heart is instrumental in generating the outgoing wave of attentional energy directed to the object."

21. See also Bradley's (2009) application of how intuition is constituted.

aries, common norms and values, and reciprocally shared positive feelings of attachment (Bradley, 2011: 302). This, once again, suggests a vibratory bio-emotional field encoded by spectra, frequency interference patterns, holographically encoding information of the group's cultural structure. Cottee and Hayward's study suggests that in response to "existential frustration"[22] engendered by diminished opportunity structures, agents find the excitement of their activities not only in line with revolutionary political goals but inherently arousing as an adrenaline rush. Thus producing both a felt solidarity with members, and "a world of exhilarating action, violence, intrigue and drama" (ibid., 979). Reviewing the biographies of other forms of combatants, they (ibid., 969) conclude that "the excitement of violence derives in part from its emotional intensity and the heightened state of consciousness that this produces." A quantized and holography version would have it that solidarity follows the dynamics of phase conjugation and harmonic resonance. Schema QD indicates that members in a more active recognition mode produce not only adrenaline rush but heightened meaning in their lives as the circuit diagram (Bergson, 2002) which they negotiate produces a flurry of recognition images that demand a new, non-conventional discursive organization. It is within this that a felt liberated "I" emerges, overcoming the existential frustration initially experienced. Terrorist identities are, however, not a pure break from a normative order, for the very new order created, a bio-emotional field, provides a bounded space from which resonance can be established with similarly minded members. It is only in resonance with this commonality that a terrorist identity and solidarity with a particular terrorist cell can be established.

Previously, we have argued that Bradley's two key orthogonal dimensions, control and flux, delineating a phase space within which the intersections are logons, quantum information, can be reconceptualized whereby "control" is an everyday, ongoing constitutive construction, and flux is a holofield. In short, the bounded field becomes a nonlinear "point of subjectification" constitutively producing terrorist identities in solidarity. As Cottee and Hayward (2011: 965; italics in original) explain, for many terrorists, it is not that they are "pushed" or "driven" by ideologies themselves, but rather "they are *drawn* to terrorism: that they allow themselves to be seduced by its own appeal; that they seek it out on their own initiative; that they are beckoned by its own peculiar magic; and that once they give themselves over to it, they find it thrilling,

22. In their words (ibid., 979), existential frustration means a "radical dissatisfaction regarding one's moral existence in the world—an emotional state marked by the feeling that one's life is meaningless, directionless, boring, banal, uneventful, anodyne, soulless, aimless, passive, cowardly."

compulsive, intoxicating—*the elevated experience of their lives*." Ideologies as attractors, function as axiomatic systems from which "truths" can be deductively constructed. The very everyday construction of ideologies (logons) of its members, following constitutive theory, draws from bits and pieces of this information (spectra) encoded within this bounded region. Further, in its construction, it energizes and stabilizes the group's identity structure.

From schema QD we can also point out the emergence of a distinct diagram (QD matrix) reflecting the place of intertwining, intersecting dyads: the ego—ego-ideal, whereby the ego-ideal reflects, to use Cottee and Hayward's conception, a circumscribed moral self[23] to which the ego refers; the ego—Other, where victims as Other are subsumed under an "it" categorization; the ego—community generalized other, where the group's ethos become much like the normative order of a street codes we found in Kubrin's study; the ego-ideal—Other, where one's own moral self with appropriate justifications or neutralizations now can downplay the thought of possible harms inflicted on the objectified other (e.g., Buber's "it"); and the ego-ideal—community generalized other where one's moral self resonates with the idealogy of the terrorist cell's. In this view, the abstract generalized other would be perceived as a boundary that needs to be opposed and crossed. From these interactions, from which a momentary matrix results, a unique signature wave is constructed which is both constantly being emitted and which interacts with incoming information-bearing frequency waves. The intensity of terrorist activity assures that emotionality will be a strong driving force. This, following Penrose and Hameroff, can culminate in the collapse of the wave function producing one distinct perception-image from the plurality of those that momentarily appear during the dynamics explained in Bergson's circuit diagram. A particular "cut" diagramed in the cross-cap momentarily produces a logon (i.e., perception-image, see Chapter 4) and an identity. A speaking subject, an "I," inserted in narrative construction becomes both the end result, but also by way of backward quantum referral, the active producer of its very identity. It is a constitutive process by which a seeming circularity reflects a process that belies strict linear causation, or strict rational choice. Logons are inherently indeterminate due to their construction in phase space and in their resulting overlapping quality described by Gabor ("overlap with the future," 1946).

---

23. Cottee and Hayward (2011: 965) have said that terrorism is both political and existential with the latter term reflecting "a set of feelings, concerns, aspirations or desires relating to one's moral self," and a choice as to "the kind of life they want to live."

6. Emergence of a Cyber-Informational-Cinematic/ Video-Driven Consumer Society

Cultural criminology (Ferrell *et al.*, 2008: 81; Young, 2011) argues that in late modernity a blurring of mass media-produced virtual images and the real is taking place where more and more the two are becoming indistinguishable. It is mediated representations that are becoming every part of reality and its construction. Images are being created from images and the process is ongoing producing a hall of mirrors (Ferrell *et al.*, ibid., Chapter 5). Baudrillard (1981, 1983) has already warned us of the emerging postmodern society (late modernity) built on image constructions and reconstructions he termed hyperreality. Ferrell *et al.* (ibid.) describe how "loops" (images of events that recreate previous images of events)[24] and "spirals" (images that become part of other images with different meanings, now reconstituted with the circulating image but never finding a resting point) together produce a social reality of crime whereby the real and the virtual cannot be separated. The meaning of crime and justice are becoming more reflections of media practices that tend toward an endless recreation, referring to its own images as the basis of yet future creations. Truth, in this dynamic is illusory, only provided support by reference to images in process of looping and spiraling. Ferrell *et al.* (ibid.) show how video games epitomize this postmodern trend. The military, for example, has embraced it as a training method; after all, youth now grow up saturated with video games and it is but a short step in converting this "training" into a military model form. More recently, Bowser (2012) has argued that the popularization by the media of hip hop and gangster rap music plays on racial stereotypes for entertainment for mass America.

A quantum holographic adaptation would stress that ever new spectra are being created in late modernity (postmodern society), a basis for phase conjugation dynamics of interacting persons reproducing the truth the images represent. In other words, ongoing momentary logons created in everyday interactions are built from the spectra, the frequency interference patterns holographically encoding information residing in the postmodern holofield. This too becomes a constitutive point of subjectification producing identities that are in tune, that resonate with these emerging images. Ontological insecurity engendered by late modernity, expressed so poignantly by cultural criminologists (Ferrell *et al.*, 2008; Young, 2011), has found its palatable correctives, amongst others, the seductive value of the flow of media images providing

24. See Manning's (1998) groundbreaking work on "media loops."

identities and packaged edgework. The internal logic of these images provides a postmodern axiomatic system which is both efficacious but yet fleeting in its spirals and loops. But this is not to despair! Deleuze and Guattari (1987) have already informed us of the process of deterritorialization an inherent logic of any assemblage and being so, can produce new territorializations, both more in a molar assemblage form, or potentially the basis of molecular assemblages, a "permanent revolution," the potential for a new order. Here, rather than the harms of reduction and repression being constitutive forces, we witness instead a becoming, ongoing mutation and transformation. These become central flows, and a person's potentials for effecting and being affected are given full reign. We shall return to this in a final chapter.

Law

Quantum holography can be applied to law to generate not only an understanding of the physicality of processes, but also to suggest how a restrictive collapse of the wave function reduces societal-wide possibility spaces. We shall briefly re-examine a Deleuzian-Bergsonian driven explanation of judges' decision-making and how judges speak in law.[25]

1. Judges' Decision-Making

Alexandre Lefebvre, in *The Image of Law: Deleuze, Bergson, Spinoza* (2008), has provided us with a refreshingly new approach in understanding how judges can be both constrained and potentially creative in their decision-making. He takes issue with classic statements by Hart, Dworkin, and Habermas as to how judges reason by "subsumption" (subsuming a case under some rule) and then are guided by deductive logic and syllogistic reasoning to a conclusion. He first examines what constitutes a case litigable in law. A situation has to be reduced to the elements delineated in law before it can be litigable. Only when a case is constructed legalistically does it have "standing" in law.

---

25. There have been some studies published in law journals (Kuttner, 2011; Kelsey, 2013; Powell and Menendian, 2010; Yager, 2013; Mootz, 2013; Porter, 1991) that develop a quantum approach to law, but as of yet, no approach has emerged which looks at both quantum and holography theory in law.

Alternatively, Lefebvre starts with some situation, generically referenced as an encounter, which in its primordial state is a complex image, a "fact-situation not yet formed" (ibid., 17; see also 58–59, 72–77). An image is defined in a special Bergsonian-Deleuzian sense as an identity of energy/material and movement (ibid., 117). It is in a vibratory, "gaseous state" and remains interconnected with all other images (Bergson, 2002: 20; Deleuze, 1986: 58–61; Lefebvre, ibid., 118). Images flow on a "plane of immanence." It is from this plane of immanence that various "cuts" of spacetime take place, "blocs of spacetime" (Deleuze, 59). An encounter must take on a particular form sufficient for there to be "legal standing." For Lefebvre (ibid., 59), "law is placed in a constitutive relationship with its outside—with cases—which forces it into action and invention." Out of the flow of primordial images that compose the universe, a special image, a "living image," evolves that has the capacity to delay and to "choose" its response. In a more everyday habitual standpoint, inattentive recognition/judgment in law assures that "facts" and law are aligned non-problematically. "Facts," that "speak for themselves," are subsumed under some legal principle and by way of deductive and syllogistic reasoning a truth can be linearly constructed. However, in the "hard cases," encounters that do not neatly fit this linear logic, real thinking must take place. Here, a "zone of indetermination" opens within the living image from which attentive recognition or judgment constructs a particular circumscribed perception of this image. This follows Bergson's "circuit diagram" (Bergson, 2002; Lefebvre, 136, 181).

Faced with a problematic encounter, the judge undergoes "leaps" into the institutional past ("pure past of the law," its accumulated past) for appropriate recognition images of otherwise primordially stored institutional memory images to find a fit with the present encounter (image). It is in this process that creative legal resolutions evolve. The decision entails an encounter, not yet formed, a primordial raw event, being compared with the "pure past of law," an "institutional being-memory." The circuit diagram provides how this process eventually instantiates a perception-image of "what we have here in law." This "cut," or "bloc of spacetime," includes some characteristics of the raw event, but excludes many others. A situation with infinite dimensions is reduced to a finite picture of what took place. Accordingly, in the "facts of the situation" normally summarized early in a judge's decision, a raw event takes on particular narrative form that is in accord with law. This is followed by the judge's use of deductive and syllogistic reasoning in arriving at a judgment, a truth in law. In the hard case where no "fit" exists, new law can be created.

Lefebvre provides an example of this application in tort law, *Palsgraf v. Long Island Railroad Co.* (1928). Palsgraf, a passenger, suffered injuries when a fireworks explosion took place on a train platform. The court over-ruled a lower court finding which based its decision on a standard of "negligence" and "proximate cause." In the *Palsgraf* decision, however, a new standard was articulated, "foreseeability." Lefebvre makes the point that the "facts" of the case, the otherwise raw event, an image, took on a new form in law as a new standard was articulated. What was verbalized in the earlier standard as to "what happened" now changed to fit the new standard. The primordial image, the raw event, in other words, was represented starkly differently. It should also be noted, had a different standard been articulated, yet a new representation of what took place (the "what happened") would have been produced.

In this rendition, Lefebvre (ibid., 158–162) follows Bergson and Deleuze's vehement disagreement with any possibility of a transcendental subject, a subject that is separate from the flow of images in a plane of immanence; here, all is one and interconnected. "Choice," as generally understood as the ability to freely choose amongst options, is not a determining factor. Rather our needs orient the perceptions we form. Implicit in the evolving perception-image is a plan of action, Deleuze's action-image (1986). "Our need," Lefebvre (ibid., 161; see also note 22, p. 275) continues, "chooses our perception; it chooses what we perceive, such that everything we do not need is subtracted away or not chosen from the image." Perception-image and action-image overlap. At the completion of the process implicating the circuit diagram, a composite perception-image is created which in itself provides ingredients for a plan of action.

Let's look at how our work might reconceptualize this development along the lines of quantum holography. We will first return to schema QD for guidance, then to a more constitutive approach in legal decision-making.

The notion of a non-transcendental subject needs revisiting, especially in light of the poignant critique of Jean Paul Sartre (1972). We remain committed to a milder version of Bergson's notion of a non-transcendental subject with the qualifications below. Our offered schema QD provides some specificity of agency. Lefebvre's model, building on Bergson and Deleuze, needs augmentation. As it stands, it is vacuous as to various aspects of agency. Recall, from our discussion on schema QD, an agent is inter- and intra-subjectively constituted.

Let's consider a pragmatic example of interacting dyads. The ego—ego-ideal dyad in law could consider Kennedy's (1977) analysis of three ideologically-oriented kinds of judges in legal decision-making. He rejects the idea that judges are merely "falsely conscious," are puppets of elites, or that they can automatically impose their ideologies in decision-making in law. He rather posits that they are, following Sartre, acting in "bad faith" and/or are making use of the psychoanalytic defense mechanism of "denial." Faced with a "hard case," the judge looks at legal arguments that are personally acceptable (this is the ego-ideal component, I(O)), and the expected or anticipated reactions by (O)others. These include: the immediate face-to-face Other, her/his inner circle of important or significant others; the community generalized other, or the "interpretive community" of legal scholars, including, generally all judges and those who research and write about the issue; and the abstract generalized other, the assumed wider societal expectations. Ultimately, s/he must abide by the duty of fidelity—doing "what the law requires." Outwardly, some syllogism needs to be followed linearly to a conclusion (truth) in law. From this, Kennedy groups his three types of judges: the "constrained activist judge" is initially guided by her/his activist agenda, but finding alternative compelling logic in law lacking, opts for a more formalistic argument (linear deductive logic); the "difference-splitting judge" looks at the extreme positions being discussed and opts for the middle course; the "bipolar judge" is a difference-splitter, making a career out of trying to line up her/his decisions evenly amongst the extreme positions, providing the sense that s/he objectively abides by the law. Kennedy highlights that both bad faith and denial are at work in order for the judge to overcome potential anxiety in decision-making particularly concerning the "hard cases." What this underscores is that Lefebvre's position in judge's decision-making needs an inter- and intra-subjective dimension. The various interactions constitute, at any moment, a matrix, a diagram of relations with particular valences. Together, they produce a unique quantum signature wave encoding holographic information reflecting this matrix of the moment. Incoming waves are perturbations that provide the impetus for new matrix construction with an altered signature wave that returns to the encounter, in constituting it in a particular form we call truth, or the facts of the case.

A second supplement already implicit in Lefebvre's position, is by way of constitutive theory and quantum holography. His notion brings out the idea of co-production. Unlike traditional law approaches where Rule X Facts = Decision, a linear process provided much initial momentum by Dean of Harvard Law School, Christopher Langdell in the third quarter of the 19th century, and much like Lefebvre's attack on the notion of decision-making by mere subsumption, a holographic inspired constitutive approach looks at how "facts" are always in a state of indeterminacy, a quantum wave, or cloud of possibilities, which can be collapsed in a particular form producing one instantiation of what happened rather than another. This collapsed form is a logon, quantum information. But the constitutive approach also suggests that this process is two way: the judge is referencing a legal quantum lexical structure in determining the "facts" in order to make it accord with a situation having legal standing, but also inserts an "I" retroactively to the initial phases as a determinant of the end result. This is backward time referral where quantum information is returned backward in time to constitute the present. The judge draws from a "pure institutional past" (rules, regulations, case law, etc.) which is a circumscribed virtual relevancy space populated by images, some of which are more "dominant memories" or "shining points round which others form a vague nebulosity" (Bergson, 2002: 171). This is a judicial quantum lexical structure, in which memory images (symbolic) are associated non-locally encoded in the holofield of the legal arena.

Returning to the ego—ego-ideal dyad of schema QD, an ego may have many orientations, one persona, or discursive subject position, in the judicial arena being a judge. Lefebvre (2008: 144) tells us the judge occupies an "institutional being-memory, the being-past of the law," where the past is a background relevancy space of rules for judgment. These are, following Bradley, spectra encoded with holographic information. Once situated in this space the available quantum lexicon is greatly circumscribed. Reality construction, the "what happened," entails drawing from this background space, making use of the symbolic memory images, guided by the effects of othering (face-to-face, community and abstract generalized others), in constructing the "what happened" with legal standing, and, in the process, clearly indicating to all formal rational objective law at work. Everyday, ongoing constructions of bits of information relate to the creation of logons which have degrees of indeterminacy, as we have previously seen. In a previous section, we have seen similar dynamics in the relation between a deprived ghetto, street codes, and gangster rap music. There, rap music is constituted from ongoing constructions of logons, which in turn relate to how given spectra (encoding street codes) are negotiated in everyday materialization.

A constitutive process is about co-production. A judge accessing spectra within the "institutional being-memory" creates a resonance pattern with her/his own unique quantum signature wave. Incoming information (from the encounter, the "hard case"), is composed of energy frequency waves that undergo a forward and reverse Fourier transform. A resonance pattern created, an informational channel opened, is the moment where, consistent with Mitchell (2008), evaluation takes place. An institutional being-memory that includes "negligence," "proximate clause" as "shining points" is also one where the quantum lexicon privileges associative connections with these notions; on the other hand, one where "foreseeability" is the shining point has a contrasting set of associative lexical connections. Said in another way, each standard is associated with a cluster of master signifiers that are the relevant ones in legal discursive production. Evaluation can take place within either loci for coherent constructions of truths, or reality.

Evaluation was not precisely explained by Mitchell, but we see that it entails engagement with Bergson's circuit diagram. We could also conceive, following Bergson, how a zone of indeterminacy opens within the living image. Or, following Penrose and Hameroff, we might conjecture that a rift or bubble in the very fabric of spacetime is created, all the way down to the Planck scale. Here, we conceive of the creation of a momentary superposition of images, quantum coherence, with a transverse plane of feelings or affection-images (see Bogue, 2003a: 146–147; Deleuze, 1989: 123–125)[26] cutting across them, a quantum tunneling, until a threshold intensity level is reached and a quantum collapse occurs (see Penrose and Hameroff, 2008). But unlike Penrose and Hameroff, this is not "objective" in the sense that no agent is actively involved. They do mention "hidden variables" at work, without substantial elaboration. Rather, our schema QD indicates that inter- and intra-subjective factors are at play in the attentive judgment mode producing a non-linear result (logons). After the collapse, or evaluation, an outward going reference wave integrating the distinct signature wave of the judge, which is always in process, returns to the emitting field, or the raw event, instantiating a 3-dimensional, greatly restricted picture of the "what happened." It would seem, outwardly to the reader of a judge's decision, and to the judge him/herself with a little assistance of bad faith and denial, that the facts and law are congruently aligned: Facts × Rule = Decision. This produces what appears as an orderly system based on formal rationality, and legislative, constitutional and case law. This further reifies the legal relevancy space, energizing spectra that encode it.

Further, we could also consider the more extreme holographic version that stems from studies of the black hole and information preservation. The legal arena is a bounded region, on whose outer surface is located, encoded in two-dimensions, the symbolic memory images of law, the "being past of the law." This is an axiomatic system, the basis of affordances that rain on the interior space and its occupiers.[27] Affordances, we previously noted are frequency patterns encoded with holographic information. Without more, they are also

26. In Deleuze's (1989: 123–125) words, "we constitute a sheet of transformation which invents a kind of transverse continuity or communication between several sheets, and weaves a network of non-localizable relations between them ... feeling is that which is in continual exchange, circulating from one sheet to another according to what transformations occur."

27. Consider, for example, the architecture of the court house, from an impressive entrance, replete with sprinklings of Latin inscriptions which few understand but think surely must reflective the sanctity of law, to the majesty of the courtroom with its decorum and ritual, to the judge's chambers suggesting a secluded place of intense objective reasoning in law.

points of subjectification, or the loci from which an eventual speaking subject is instantiated.[28] But, unlike Gibson's notion of consciousness being about "direct perception," linear points of subjectification, we argue that schema QD provides the framework, or diagram, whereby nonlinear dynamics and an inherent indeterminacy, particularly as to how a quantum wave is collapsed and how logons are situated, lead to creativity as well as reification. The speaking subject, "I," is both constituted by, as well as constituting via backward time referral, the emerging perception-image.

Lefebvre's groundbreaking working on applying a Bergsonian-Deleuzian perspective, however, does not address how it is that a subject (judge) speaks. Integrating quantum holography and constitutive theory with Lefebvre's work on how a judge decides clarifies the physicality of how the "out there," the raw event, is constructed as a case with legal standing.

2. Judges as Speakings

To briefly address the issue of how the judge speaks, we first find Lacan's (1977: 306) "graphs of desire" illuminating for a composite statement of how speaking takes places. Previously we summarized his "graph 2" that shows how an anticipatory, forward time dimensions combines with a backward time dimension ("retrograde") in producing each signifier in a signifying chain ("punctuation"), and how, via various feedback loops (imaginary and symbolic), coherent narratives are created.[29] His model is based on an oedipalized notion of a lacking subject, the cost for entering the symbolic order which creates both mastery and a castration (separation) from the real. We, alternatively, view this form of lack induced by political economy (see Deleuze and Guattari's *Anti-Oedipus*, 1983), and that a more primordial ontological foundation rests on the notion of "production" (Deleuze and Guattari, 1987), conceptualized as a perpetual becoming, a drive toward increasing one's capacity to effect and be affected (Spinoza, 1985; see also Hardt, 1993: 56–111), a point expressed earlier by Alfred Whitehead (1925, 1929). Bergson's notion of "élan vital" and Freud's notion of the libido are relevant for accommodation.

In Lacan's (1977: 313) graph 3 of desire, a more complex diagram, a key component is when some question is addressed to the Other (recall: "che vuoi?,"

28. All know their place, or discursive subject position, when entering a courtroom, variances to which will be met by contempt of court citation.

29. Thus, we could diagram this as: S/s-s-s-s-s. → s(O), or S/s. In quantum notation: $|s_i>$. A particular signified is collapsed from the possible signifieds.

what does it want from me?). This is in accord with Bergson's internally directed questions, and with Stapp's Process 1. We invoke the circuit diagram at this point, and perhaps Penrose and Hameroff in defining the point at which the collapse of the wavefunction takes place. In his graphs, Lacan shows how signifiers come to be attached to particular signifieds, or: "punctuation," s(O). Signifieds otherwise "slide" under the signifier, much like quantum theory would suggest (i.e., a "cloud of possibilities," or $|s_i\rangle$). More recent work, prefaced with the idea that "language is physical" (Benioff, 2003), provides quantum holographic principles that can be integrated with Lacan's de-oedipalized graphs of desire. Chapter 4 summarized key points in a quantum models of the human lexicon (Plate, 1995, 2003; Jones and Mewhart, 2007; Cox *et al.*, 2011; Hannagan *et al.*, 2010, 2011), word associations and recall models (Bruza *et al.*, 2008; Bruza *et al.*, 2009a, 2009b), and a phase conjugation model (Marcer and Schempp, 1997, 1999; Mitchell and Staretz, 2011). There is much work to be done here. We only make several forays into this inviting area.

Let us provide a first approximation. Once again, the work of Deleuze and Guattari are suggestive for developing a composite picture. Judges work in a particular bounded region of law which offers distinct personae (judges, prosecutors, defense counselors, etc.) or better discursive subject positions, and a "pure past of law" that is a repository of legally constituted memory images, a bounded semantic and syntax structure from which elements for the construction of legal reality can take form. This totality, as well as other components of this "law machine" (Bogue, 2003b: 78),[30] can be represented as an assemblage(s), that has an operative abstract machine acting much like a pilot wave which is primarily the logic of formal rationality, linear deductive logic, and syllogistic reasoning. The machine operates by ongoing "cuts" of quantum flows in the plane of immanence. This bounded space which is a field of "virtual vectors," or for us, legally proscribed affordances and "order words," provides the background relevancy space for recurring forms of social interactions, discursivity, and constructions of legal reality. Marcer and Schempp (1997, 1999) as well as Mitchell and Staretz (2011), have shown (recall from Chapter 4) how holographically encoded raw material "maps" to the "holographic encoding of the arbitrary label for that object." We conceive this to be saying that signifieds representing images are attached to particular signifiers. By "mapping to" they mean phase conjugation. Accessing a signifier also places

30. Bogue (2003b) follows Deleuze and Guattari's examination of Kafka's *The Trial*. The law machine is a "social machine whose components are people, texts, institutions, practices, buildings, objects, and so on" (ibid., 78).

one in a resonating dynamic with a signified, the latter being in flux until a collapse of the wavefunction produces one signified from the plurality of possibilities, or s(O).

Drawing from recent work in quantum and holography theory concerning human lexicon we also see that the field of law offers associative word clusters which are accessed non-locally ("spooky activation at a distance"). "Negligence," "proximate cause" or "foreseeability" as a standard in civil law, for example, operate as a loci of "shining points," or associative-word complexes/ matrices. By accessing one matrix or complex rather than another, particular complexes of associated words (signifiers) are activated which resonate with particular signifieds. Proximate cause, for example accesses an associative complex of signifiers that are connected with the notion of more immediate effects of the defendant's behavior. Lefebvre (2008: 125), for example, notes that many factors in the *Palsgraf* decision were not mentioned in the legal construction but were mentioned in the initial briefs, such as whether the blast itself or the crowd that was panicked was the cause of the overhead scale falling on the plaintiff. The standard of "foreseeability," on the other hand, accesses an associative complex of signifiers that would focus on whether a small package carried by a commuter could be dangerous and could have eventually produced the injury to the plaintiff, Palsgraf.[31] Thus the various "cuts," or blocs of spacetime engendered within the boundaries of the law machine already are limited by the existing relevancy structure.[32] This can be conceptualized in the more inattentive judgment mode, as a more linear point of subjectification. Constitutive theory would then suggest that more dominant understandings of reality are continuously reified.

31. Thus in post 9-11, standards for judicial intervention in creating frameworks for grounds for surveillance have exponentially expanded. This of course, places a considerable larger segment of the population under surveillance, as we have seen with the recent U.S. massive surveillance under N.S.A., and disclosures brought about by Edward Snowden.

32. The U.S. criminal justice system can be viewed as offering greater excusability as one moves up the socioeconomic scales, and conversely, placing lower segments at risk in so much as their indigenous associative word networks are greatly limited for inclusion in crime discourse. Consider a lower class shoplifter who attempts to widen his defensive discursive explanation by including poverty, unemployment, discrimination, etc. Now consider upper echelons in many injunction proceedings who need to show "why is it not the case you should be stopped in doing what you are doing"? If we had genuine formal equality, why shouldn't the shoplifter be given the same opportunity in drawing from a wider associative word network? Would the political economic system be in danger of facing its internal contradictions, on the one hand, and on the other, producing alternative discourses that provide the potential for more critical, system-destabilizing narrative constructions?

Schema QD informs us that the speaking subject (Lacan's "l'être parlant," or "parlêtre") situates him/herself within this bounded region of signifiers and operative rules or codes for stringing them together in discursive production. Thus when a judge is faced with some raw event, an encounter, which is a "hard case," s/he engages not only the signifiers available from the holofield of this bounded spacetime but also, from schema QD the intensities and instantiations of the three others, as well as demands to be true to an ideal conception of self (ego-ideal). The raw event, or pure image must be constituted as a case in law as we have seen, but also must be expressed in discourse. Lacan's graphs of desire are suggestive in showing how signifiers that are more wave functions, encoded with holographic information, undergo particular instantiations in the collapse of the wave function. Phase conjugation dynamics indicate how signifiers accessed from the legally constituted holofield come to resonate with a particular signified (Lacan's punctuation point). In the hard case, with activation of attentive judgment, the circuit diagram indicates how several versions of signifieds are constructed and exist side by side in a state of quantum superposition. Various signified, in other words, are superposed, each a possible instantiation.

In a forward direction, each signifier in a state of being completed anticipates, due to overlapping of the sinusoid, the next, as Gabor's (1946) study of logons as quanta of information suggests. At a threshold, the collapse of the wave function instantiates one signified rather than another (punctuation point). The judge's unique signature wave reflecting a momentarily fixed schema QD matrix now returns resonating with the incoming object wave creating interference patterns, a connected series of signifiers, at the end of which, Lacan's pause, provides retroactively (a "retrograde," or for us, backward time referral) the distinct meaning of the sequence of signifiers. The judge as speaking subject, "I," is also shaped by his ideal-ego construction (consider Kennedy's three kinds of judges). In short, the bounded sphere of law, the law machine, emits affordances or order words, which become the basis of points of subjectification. The judge, as speaking subject, is both author via backward time referral of quantum information, and object in this process. In the inattentive judgment mode, we do foresee more linear constructions of both the raw event as a case with standing and the judge with a particular ideological persuasion; in the attentive judgment mode, we concur with Lefebvre's analysis in showing that engendered are nonlinear constructions that offer creative alternative in these construction, e.g., consider the development of the Miranda warning, right to privacy, affirmative action, "native title," etc., that did not follow linearly in law. In the next chapter we will explore how non-reifying instantiations can become more ubiquitous in a socius.

Conclusion

This chapter has been about applications of quantum holographic principles in criminology and law. We provided suggestions for future lines of flight that will take this material to the limit and beyond. We make no pretenses of making a comprehensive analysis. We merely want to show that alternative concepts to the traditional paradigm provide a new beginning. There clearly is much to be done. We have also seen that a number of contemporary works can be augmented by further incorporation of a process-informational paradigm. Cultural criminology is provocative and insightful, although its statement on agency is in need of development. Studies on gangster rap music and its relationship to violence have provided much stimulating and informative analysis of the relationship between deprived areas, street codes, gangster rap music, and violence. This we saw was not a linear relationship, but is constitutively produced. Legal scholarship has been greatly benefited by Lefebvre's study of judge's decision-making, making the case that the "good profit" (Rorty, 1989) has a place in the attentive judgment mode of orientation. In the next chapter we will have further thoughts on the potentials for a transformative justice, a necessary replacement to merely restorative justice. Our schema QD, no doubt, will be the basis of much critique and commentary; but to the degree it brings the subject back in and furthers a quantum holographic based identity structure, it will make our initiative worthy.

Chapter 6

Transformative Justice

Introduction

Our criminal justice system is a failure, pure and simple. If anything, it reproduces, stabilizes, and perpetuates criminal identities. We must move beyond; we must exercise what C.W. Mills (1959) and Jock Young (2011) to follow coined a "sociological" or "criminological imagination." We need not rehearse the harms of reduction and repression ubiquitous in contemporary political economy. They are plainly too numerous to rehearse. We can cite race, sex, gender, intersectional harms, as well as extreme inequalities in wealth and access to resources; we can cite the exclusionary society, uncertainties and instabilities of postmodernism/late modernity; forms of othering that objectify and reduce; forces of rationalization and commodification that render the person an object within an overall abstract machine; and the emergence of a disembodied, media-driven, virtual-oriented culture. More directly, we only note in passing the 2.4 million people within the prisons of the United States, with an assured outstanding two-thirds recidivism rate after three years. What can be done? We have recently seen movement toward restorative justice as a more inclusive approach where participants are active in justice rendering. We have also noted the criticism of this approach in its dangers to being subsumed within "criminal justice." We opt for a transformative justice. It has not yet been adequately formulated. Here, we continue our ongoing development of approximations to a desirable orientating model. Transformative justice is a holistic approach, prioritizing policies and programs that include interactions between all "levels" of society, from the more macro- to the more micro-level. Quantum holography is well situated as an alternative, a past-due ontology, that along with progressive theorists, most noteworthy Deleuze and Guattari, and those who have applied their works, such as Hardt (1993), Hardt and Negri (2001, 2005, 2009) and Holland (2011), provides the basis for rethinking genuine change at the person, community, and societal levels.

We have seen that integrating quantum holographic concepts with the work of Deleuze and Guattari offers a new beginning in the social sciences, and more particularly in criminology, law and notions of a transformative justice. Normal science based on Newtonian physics provides a static view of the universe. We have recently witnessed the beginnings of a more process-informational paradigm.[1] Here, we will address three critical components connected with a transformative justice: molecular assemblages, discursive reconstruction, and evaluative schemas and ethics.

Toward a Permanent Revolution: Molecular Assemblages

In Chapter 3, we reviewed a socius as being a historically situated, relatively autonomous, ubiquitously penetrating, dominant assemblage, continuously reified by agents interactively producing a social reality by assuming non-examined ontological assumptions. It is further composed of a multitude of bounded and overlapping assemblages extending from the more "macro" or

1. Deleuze and Guattari, in their late work, *What is Philosophy?*, argue that we need to cultivate asking the right questions and to develop novel concepts. Woolford, in *The Politics of Restorative Justice*, argues that we must deconstruct traditional identities and replace them with more dynamic, open, mutating forms and a transformative justice as "an ongoing project" guided but not constrained by new ethical principles. Dyer-Witherford, in *Cyber-Marx*, has provided us with a provocative analysis of the internet revolution in relation to political economy and concludes with a call not for a blueprint, but a "battlefield map," "series of initiatives whose advancement would contaminate and overload the circuitry of capital with demands and requirements contradictory to the imperatives of profit ... and increasingly require the reassembly of everyday activities into a new configuration." See also Castells (2012) and Benkler (2006) who show the transformative potentials of the internet. In line with this, Unger in *False Necessity*, argues for role jumbling and boundary crossing, the logic of which can be further enriched by Foucault's notion of embracing "limit-experiences, a form of edgework, but rather than adrenaline rush associated with approaching boundaries, suggested is in fact crossing boundaries (See Lyng's clear explication, 2005: 39–47; see also Ronnie Lippen's notion of a boundary crossing criminology, 1998). Holland, in *Nomad Citizenship* has provided a societal model based on the notion of Deleuze and Guattari's *nomadology*, citizenship based on a form of being a jazz player. Hardt and Negri, in *Multitude*, have offered new forms of collective social action that deny stasis. Grosz, in *Volatile Bodies*, Cornell in *At The Heart of Freedom*, and Butler, in *Bodies that Matter*, amongst many emerging views along these lines are calling for a "refiguring of bodies" and safeguarding the imaginary domain. The emerging field of cultural criminology, as articulated by Ferrell *et al* (2008) and Young (2011), offers a political economy of present day cyber society and notions of oppositional forms of edgework. And ultimately, we are receptive to the call of Deleuze and Guattari (1987) for "permanent revolution."

institutional level across every "level" to the more "micro" level. These, as we previously summarized, following one version of the quantum holographic theory (e.g., Susskind, Boussa, 't Hooft, Maldacena, Bekenstein), can be seen as bounded regions with all information within the bulk encoded on the surface boundary. The socius, then, includes the particular axiomatic system crystallized at a particular moment in history. An axiomatic system, following Deleuze and Guattari, is a system of capture, a systematic way of creating "cuts" in the plane of immanence, blocs of spacetime. Thus, an axiomatic system offers key truths as major premises that work their way via deductive logic and syllogistic reasoning to resonating truths created within the bulk, within everyday social activity. We have seen how Foucault's disciplinary mechanisms and capital logic are abstract machines, and, following Gibson, the origins of affordances that are negotiated by agents conceptualized in terms of the dynamics of schema QD in producing particular logons, or perception-images. Phase conjugation and the collapse of the wave function are involved in a complex process of virtual image construction and development of a speaking subject, particularly in the active judgment/recognition form of agency and subsequent motor activity. Disruption in the process of more habitual reality construction is the key, as Bergson has shown us in his classic *Matter and Memory*, for creative evolution, mutation, and transformation. So we arrive at a non-deterministic model that acknowledges both an objectification process, manufacturing forms of othering that are "it" oriented, and the ever-potential reality of forms of othering recognizing becoming and the "thou."

1. Subjectification and Deterritorialization

We are further aided by Deleuze and Guattari's notion of deterritorialization. All assemblages, we previously summarized (see Chapter 3), have an inherent destabilizing element that subverts that which is in process. More dominant assemblages or strata according to Deleuze and Guattari (1987), or what they referred to as molar assemblages, are static, driven by homeostatic dynamics of repetition of the same. It is the system of capture of otherwise flows of clouds of possibility, best mathematically modeled by the Schrödinger equations that collapses the wave function into a more static, immobile form on the plane of immanence. An embedded axiomatic system with its abstract machine, such as capital logic, panopticism, formal rationality, linear deductive logic, and "normal science" generates affordances that traverse the socius and the various assemblages within it. More dominant affordances can be conceived as forms of solitons, or quasi-pilot waves that not only traverse the spacetime bulk, but put in resonance disproportional amount of spectra within this bulk.[2] Since all entities or "appearances" within the enclosed energized plane of immanence, or spacetime encode all information about anything that has ever happened to them, we argue that the historical result is differential distributions of reactive and active informational-bearing spectra. Reactive spectra are highly encoded with information that tend toward repetition of the same, ossification, and stases (i.e., more molar in form); active spectra, on the other hand, encode information that reflects more open, dynamic, novel, mutating clouds of possibilities (e.g., more molecular in form). With the collapse of the wave function, the former reflects a more limited range, the latter, a greater range of possibilities.

The axiomatic system is the point of subjectification from which a speaking subject is created, and who, in the linear mode of inattentive recognition/judgment, reifies it. It also reflects an associative lexicon, or relatively stabilized loci or matrices of interconnected signifiers pointing to the sense of the word in context (see Kitto *et al.*, 2013; Kitto and Bruza, 2011).[3] This sense of a word in context is encoded in a "conceptual space" (Gardenfors, 2004; De Vine and Bruza, 2010; Kitto *et al.*, 2013), a background space of relevancies accessed by those with whom phase conjugation produces resonance and informational flows. Spectra in the holofield are continuously re-energized in use, even as everyday logons, products of the interactional dynamics of schema QD, provide some reprieve from linear reconstructions. We also noted that in a more attentive recognition form, a suspension in linear creations and reifications of "reality" takes place, during which time a zone of indeterminacy opens, a space in which various recognition images become superpositioned, and, when a threshold is reached, collapse into one, which becomes the virtual image emitted outward toward the object, or encounter, in constituting it in a three-dimensional form. This is more apparent in molecular forms of assemblages. Molar assemblage's expressive and constitutive forms are more in line with Gibson's notion of affordances and "direct perception," a form of inattentive recognition and the recreation of the same.

Molecular assemblages are characterized in dynamic systems theory more as dissipative structures which provide for both a relatively stabilized basis of action, but yet remain sensitive to perturbation. They are disproportionally sensitive to initial conditions and operate by nonlinear dynamics, as dynamic systems theory clearly shows. Their natural home is far-from-equilibrium con-

---

2. That is, consisting of the spacetime, holofield, plane of immanence, or ZPF.

3. Kitto *et al* (2013) retain a quantum approach as to how information is stored and retrieved. Vine and Bruza (2010), however, do open up the possibility that their model, an expansion of Kitto *et al* (2010), could include holographic theory.

ditions. Here, not point and cyclic attractors are privileged, but torus and strange attractors which both are ordered and dis-ordered at the same time. If, as contemporary social critics suggest, late modernity systematically engenders harms of reduction and harms of repression, and, in response, tendencies to ossify molar forms of assemblages (Young's late modernity and forms of othering and identity politics), then, change must entail the transformation from molar assemblages to molecular, and harms of reduction and repression must be countered by a converse more liberatory ethic. It is within relatively autonomous bounded fields of predominantly molecular assemblages that active spectra will become disproportionally embedded. A person, therefore, within this spacetime would have available for resonance more active forces (spectra) paving the way to becoming-minor, an improvisational jazz player. Each momentarily created logon, consistent with Bradly's application of Gabor's (1946) quantum holography approach, would have degrees of indeterminacy opening space for greater choice, especially of active spectra. The speaking subject, "I," that is both the producer and produced result of this process would find greater power in her/his capacities to both effect and be affected, as we will show in more detail below.

Molecular assemblages tend to constructions of lines of flight that are constantly deterritorializing and reterritorializing, a process of becoming, a becoming-minor, becoming-other and tending toward the development of a minor discourse (Rodowick, 1997: 156). Initially, commonalities will appear as a rhythm or tune, a "refrain," and subsequently, with "more play," assemblages (Deleuze and Guattari, 1987: 313). This is a common resonating wave, a "unique signature of group affiliation" (Bradley, 2011: 198), or what Wendt (2007) calls a "social wavefunction," that is encoded as a hologram "in the energy spectra of vocal interactions among group members." It can reflect either a more community generalized or abstract generalized other. Expressive forms of this field and its molecular assemblages would appear, in one form, as a minor discourse, a discourse composed of signifiers that resist closure (more wave-form than particle-form), that are constantly being subject to mutation and metamorphosis (see also Mitchell and Staretz's model of social change, 2011: 14; Bradley, 1998: 131–132).[4] Here, not "becoming-the-same" but

4. For Mitchell and Staretz (2011), each interaction with an entity produces a phase shift in its hologram. After resonance (phase conjugation) is established, an evaluation takes place by the receiving entity, which, in our view now implicates both Bergson's circuit diagram, during which a number of memory images (recognition images) are sought and matched with the incoming wave frequency, and the various inter- and intra-subjective components of schema QD—together they tend to produce a distinct wave reflective of

becoming-other, an identity constituted by inter-subjective recognition and appreciation of the Other, would be the basis of a people who are not yet, but in process, becoming (Deleuze and Guattari, 1987; see also Rodowick, 1997: 154–156). Schema QD's others, in other words, would not be static, but becomings. Consider in this context, for example, Levinas' (1998) much cited notion of an infinite duty of care to the Other, an Other that cannot be bound in rigid categorical distinctions. A "collective enunciation only takes place through a minority language where I becomes other and others can begin to become in recognition of a collective elaborate I" (Rodowick, 156). To this end, Deleuze's (1986, 1989) cinema books suggest strategies "for mapping becoming without immobilizing it" (ibid.). "Becoming must keep on becoming, in an indefinite movement of invention opening wider and wider zones of autonomy populated by more and more singularities ..." (Massumi cited in Rodowick, 156). Specific strategies suggested include: creolization, or a deterritorializing of dominant languages, creating a new language from within; "fabulation," or telling stories that resurrect denied voices; and the use of intercessors, a process by which the cultural revolutionary lives a script of the disenfranchised indigenous person, who, in turn, plays out her/his role as change agent, both in process being an educator and educated. This, Rodowick (ibid., 156) says, is not to attempt to represent a people,[5] but rather, "an idea or image must be constructed through which a people can affirm their becoming as a collective will to power." This is in accord with our alternative paradigm being developed. Privileged is the dynamism of the wave-form over the static particle-form, sensitivity to initial conditions, and nonlinear dynamics, over objectifying and rigid formal rationality.

Bradley's (1998, 2010, 2011) discussion of the relationship between flux (fluctuating energy in movement, encoding holographic information and its activation and expenditure) and control ("a hierarchical order ... a highly stratified structure of asymmetrical relations ... a system of social constraints,") which "directs the expenditure of energy," (Bradley, 1998: 128, 130), and how everyday collaborative behavior of persons coordinates expenditure of energy, together act as a communication system that is responsible for the production of Gabor's (1946) "quantized units of information." Flux and control, he

---

the various states traversed. At a threshold value, after a transverse wave (quantum tunneling) representing passions and the emotions cuts across them all, the superpositioned recognition images collapse into one that becomes representative of the incoming object wave and then is emitted outward toward the object constituting it in its three dimensional form.

5. This is the basis of an antirepresentationalism we will come across in discussing a retributive ethics below.

Figure 4. Communication, Organization, and Logon Construction

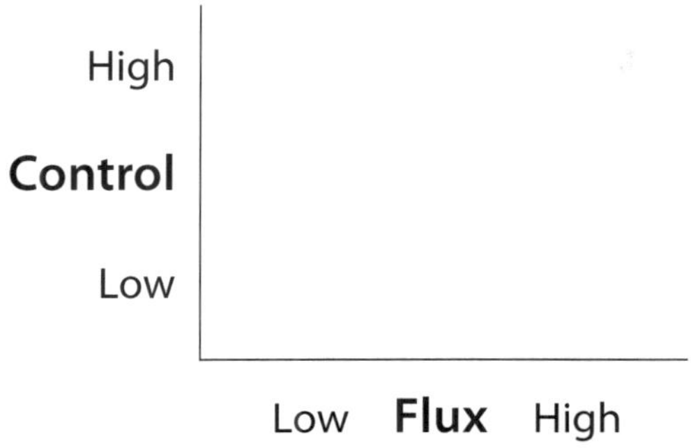

Source: Adapted from Bradley, 1998: 134, Figure 4.

shows, can be mapped as two (orthcgonal) dimensions on a phase map ranging each from low to high (Bradley, 1998: 133–136) producing characteristic attractor states.

Thus low control and high flux produces "volatility" (e.g., lack of hierarchical control leaves energy flow in amorphous, fluctuating states); high control and low flux produce "ossification" (e.g., rigid forms of organizations such as bureaucracies); low and low produce "insufficiency" (i.e., little basis for coherent organization and information development); high control and flux produces a "transformational and novel" state; a middle position of flux and control is said to produce a "functional and stable" state; in between the latter two is an area of "turbulence" and "instability" (e.g., where nonlinear dynamics may push the more "functional and stable" state into a transformational and novel state). This is a form of symmetry-breaking found at bifurcation moments in the model (Bradley, 1998: 135; Delanda, 2002: 18–21). Nonlinear dynamics, such as the logic of small inputs can produce disproportional effects, can effect a phase shift from instabilities to stabilities. The model also suggests that outside of the functional and stable state, perturbations in flux and control can push the otherwise unstable states into the former at critical threshold levels. We have found guidance as to this movement in Deleuze and Guattari's (1987) notion of the refrain, a tune, or melody that resonates, producing, with greater intensities, assemblages, and under an axiomatic system, molar assemblages. Bradley (ibid., 135) concludes that "the fluctuations in flux and control crystallize into an emergent order of social interdependence signified, normatively, by the creation of a membership boundary." In other words, most efficient information processing is created within the stable attractor state (ibid., 136). But stability, Bradley continues, requires both a threshold level of flux and control and a linkage or "coupling" amongst members of a group (or socius) and the specific form of hierarchical control. This is a "network of socio-affective relations" or "bio-emotional relations" which can be conceived as encoded in a distinct group wave frequency, or "identity signature" (Bradley, 2011). It is with this that members form phase conjugation processes, forward and reverse Fourier transforms, and resonance—all tempered by an agent seen in terms of our schema QD—in creating the group (cultural) construction of reality, a common sense notion of the wherewithal of "reality." Here "information about the features and organization of a whole (an object or an event) is spectrally encoded into the oscillations of energy as an interference pattern and is distributed to all points and locations throughout the field" (Bradley, 2011: 212; Gabor, 1948). They reflect the form the community and abstract generalized other have taken. These shape affordances, and in the inattentive recognition/judgment orientation, direct perceptions result in reification.

To further bring this discussion within quantum holography and the work of Deleuze and Guattari, we could conceptualize "control"—a normative, cultural order, with language as a central element—as high control indicating molar forms of language, or major-literature/discourse, constituted by "order words" (Deleuze and Guattari, 1987), "master signifiers" (Lacan, 1991), and less variability (Deleuze and Guattari, 1987); whereas low control indicates minor-literature, continuous variation (Deleuze and Guattari, 1987) and processual forms of master signifiers (Bracher, 1988, 1993).[6] To integrate Deleuze's thoughts from his cinema books (1986, 1989), the former would be more akin to the organic regime of signs and privileging of movement-images producing more linear social reality constructions, discovery of assumed pre-given objective realities, and in the extreme, "the man of truth"; the latter, more crystalline regime of signs, and privileging the time-image, the ubiquitous emergence of falsifying narratives, and ongoing creative constructions of truth. Modernity is about the predominance of the organic regime of signs; late modernity, as envisioned by Young (2011), or better, postmodernity, finds a crystalline regime of signs more prevalent. For the latter, things are more in flux, in a state of perpetual decoding, mutation, metamorphosis, becoming.

6. Bracher (1991: 72–73) tells us alternative forms of master signifiers "are less exclusive, restrictive, and conflictual [they are of] a different style ... a style that is less absolute, exclusive, and rigid in its establishment of the subject's identity, and more open, fluid, and processual: constituted, in a word, by relativity and textuality." For an application of the four discourses to law, see Milovanovic (2004) and Schroeder (2010).

Looking at the two orthogonal dimensions, flux and control as delineating a phase space, would also suggest various basins of attractions, i.e., the various loci of their intersections. Each locus, or region, in turn, can be conceptualized as a vector space, consistent with the notion of Hilbert space, where each vector represents multiple values in movement in spacetime. These loci are also constituted as a semantic space (Widdows, 2004; Bruza and Cole, 2006; Kitto *et al.*, 2013) reflecting a circumscribed quantum holographic lexicon,[7] in the form of matrices of associated words (De Vine and Bruza, 2010; Jones and Mewhort, 2007; Bruza *et al.*, 2008; Hannagan *et al.*, 2012). Here, each vector represents both various possible values (meanings) of words *and* context, simultaneously (De Vine and Bruza, 2010; Bruza and Cole, 2006). In quantum notation, the "basis" state is written: $|w_i>$. It is the collapse of the wave function that instantiates one particular meaning of a word in a particular context, and, as we have seen, further subject to the particular form that schema QD takes. Thus, we could take a particular attractor basin, say located at the intersections of low control and high flux, and identify it as a locus of word associations existing in a semantic space, a circumscribed vector space reflecting potentialities in word values. This is akin to what Bergson over one hundred years ago referred to as "shining points" or "points of luminosity." Words "chosen" within these loci exist in association with particular other words, as research in both quantum and holographic approaches to human lexicon show (Gabor, 1968; Longuet-Higgens, 1968; Poggio, 1973; Plate, 1995; Cox *et al.*, 2011; Kelly *et al.*, 2013). That is, when one is chosen, it is more likely to bring forth other associated words by way of non-locality (i.e., a "spooky action at a distance" suggested by Bruza *et al.*, 2009). The collapse of the wave function reducing the basis state representing a word's potentials (coded meaning *and* context) takes place in particular contexts of use. Thus in the high control, low flux, we suggest that the collapse of the wave function produces narrow instantiations of possibilities.

We can therefore conclude, that each locus on Bradley's two-dimensional phase space also represents a "point of subjectification" (Deleuze and Guattari, 1987). Thus the diagram indicates loci of subjectification. That is, these are the loci from which, as Deleuze and Guattari (1987) have pointed out, a speaking subject emerges. In the high control, low flux, ossification means that there exists rigidly connected word associations that resist change, and are the loci of subjectification of speaking subjects (i.e., "I") that constitutively reconstitute

7. Bruza *et al* (2009: 2) informs us that "a mental lexicon refers to the words that comprise a language, and its structure is defined here by the associative links that bind this vocabulary together."

and reify the dominant order. In the low to medium control, particularly as we move toward the low, by high flux loci we argue that an emergent crystalline regime of signs is more akin to a becoming-minor of literature and discourse; in other words, associative word linkages are more dynamic, changing, mutating, and in a process of metamorphosis. To return to Young (2011), late modernity, or postmodernity is not necessarily grounds for despair, for it is also a field in a perpetual state of decoding, flux, recoding, reformulation. Persons in this locus, subject to the dynamics of the loci of subjectification tempered by the dynamics of schema QD, are offered the opportunity to engage in becoming-other, becoming-minortarian. The potential is there, it only awaits practices that are along these directions, rather than practices of despair, loss of hope, pessimism, and forms of othering that distance, categorize, reduce, and repress.

Let us provide a concrete example, drawing from Bogue's (2003b) reformulation of the interpretive work by Deleuze and Guattari in *Kafka: Toward a Minor Literature* (1986). Bogue argues that in bureaucratic forms (referencing the Austro-Hungarian Empire), an existing assemblage can be seen as a "virtual plane of 'lines of flight' " which represents "tendencies," and functions as a "pilot role"; in short, a form of an abstract machine (ibid., 80–86). These vectors, because of their inherent potentials, are said to be "immanent" within bureaucratic assemblages and potentially serve a "pilot role for metamorphosis." However, one cannot determine in advance, Bogue instructs us, how vectors of deterritorialization and reterritorialization take specific forms (ibid., 84). Remaining in potential, then, are always lines of flight leading to revolutionary form of metamorphosis (ibid., 84–85). (We will return to how these can be evaluated, below.) Bogue does suggest "actualizations" in this direction in his practice of "active dismantling" based on the immanent virtual lines of flight that always provide potentials for being otherwise.[8] We can see that these vectors are quantum wave functions encoding holographic information. And it is the collapse of these wave functions in combination with phase conjugation that alternative information is instantiated. (We shall return below to de-

8. Bogue (2003b: 81) referencing Kafka's *The Trial*, informs us that in "Kafka's transcription of the Law as a desiring machine is also a dismantling of the machine, an analysis of the ways in which the forces of domination and authority are undone, reconstituted and redeployed in unpredictable configurations." Elsewhere, we (Milovanovic, 2004) have suggested that a reinterpretation of Lopez's *Rebellious Lawyering*(1992) along the lines of dynamic systems theory in combination with Lacan's four discourses and Paulo Freire's dialogical pedagogical approach suggests that new forms of master signifiers, "true words" could be developed that lead to alternative forms of social reality constructions. For another day, we could integrate quantum holography to provide the physicality and specifications of the process of this information construction and transmission.

veloping a statement of ethics and an evaluation schema in determining which forms, positive and negative, are instantiated.)

Bradley's (1998) comparison of more traditional music compared to jazz is instructive. The musical score is a form of "control" instructing participants how flux can be activated (a tune) within certain parameters (i.e., beat, pitch). We shall see below that Holland (2011) presses this point to show how an orchestra is an hierarchically structured organization where flux has only specific ranges of actualization (i.e., performance). Bradley (1998: 132–133) does consider jazz players, but notes there is at least a minimum control to which they must be responsive; that is, at least the "key" and time/beat variations in playing a key in concert with others must be subject to some form or regulation.

Bradley also instructs us that in the highest form of control and flux conditions, much effort must be expended to maintain stability and that "sudden organizational spasms" could occur in which case either a structural transformation takes place, or a "devolution," a "complete breakdown and collapse of the collective as a viable organization" (ibid., 136). He provides some empirical results in his study of various forms of organization in communes (ibid., 137).

This model can, however, benefit from a reformulated notion of control, which in our view would more closely follow dynamic system and constitutive theory.[9] Holland (2011: 20–24) offers an alternative form of organization based on "nomad music" such as in improvisational jazz compared to an orchestra, and soccer compared to North American football; and thus, football in this sense is more like an orchestra, jazz more like soccer. In Bradley's model, this would be an attractor basin where low control intersects with high flux, or the attractor basis of "volatility." Here, control is constitutively constructed in usage. And it is precisely here that Deleuze and Guattari's work on deterritorialization is appropriate for consideration toward a transformed society (socius). In this reformulation, "control," can be loosened to engender destabilizing (deterritorializing) lines of flight, decoding of the social field, and their ongoing reconstruction (reterritorialization) into more molecular forms of assemblages. This tends toward a state of "permanent revolution."[10] In this alternative space, logons constructed would be reflective of a more expansive range of possibilities and varieties of instantiations, a becoming-minor. Expressive forms, particularly discourse, could begin to become minor literature

9. Here forms of control are reconstituted in everyday life by the inadvertent or more unconscious usage of primal system-maintaining elements, as is the case in phase conjugation with a more pervasively accepted and accessed conceptual spaces populated by holograms. See also our earlier discussion of gangster rap music (Chapter 5).

10. For "permanent revolution," see Holland, Chapter 4 (1999).

(molecular form), following the dynamics of the "refrain" (Deleuze and Guattari, 1987) whereby new resonating frequency patterns take more stable form (holograms) in the eventual development of molecular assemblages. The agent who engages this field would also be more a becoming-other than a becoming-the-same. Murray (2013: 48–52), for example, has provided an insightful notion of a "vagabond thought" that would be actualized within this emerging molecular space. It is characterized as a "wandering," a resonating ambivalence, and "continuous variation" (Deleuze and Guattar, 1987). This mode of thought, a hybridity, triggers novel incorporeal transformations, much like that suggested by Foucault's notion of "limit experiences" (see Lyng, 2005: 45–47), a transgression of boundaries, a "border-crossing" (Lippens, 1998).

Late capitalism, according to Deleuze and Guattari (1987: see also Patton, 2000), is characterized by a perpetual system of decoding and recoding. It has provided flux myriad forms of expression compared to previous forms of societies hindered for example by the rigid constraints of tradition, religion or culture. However, as flux is provided greater forms of expression, as molecular assemblages take form, they are brought within the axiomatic system by way of a system of capture (such as in commodity, legal, and linguistic fetishism, media practices, consumerism). But even as capital logic tends towards ubiquitous influence in political economy, tending, in the extreme to high flux and high control—i.e., axiomatic, circumscribed order words and affordances, abstract machines such as in deductive/syllogistic reasoning, normal science/paradigms, and organic regime of signs that provide a linear model of instantiating the wave function—there remains, as Bradley's model predicts, an opportunity of a "spasm" or singularity or bifurcation point to emerge, sending this dynamic system toward a phase space in which transformation and novelty materializes, or to devolution, a structural collapse, as Marx envisions the destiny of late capitalism, a qualitative change to a new socio-economic order.

2. Reconstructive Nomadology

Recent works by Hardt and Negri (2001, 2005, 2009) on the "multitude" and a new commonwealth that are based on Deleuze and Guattari's ontology provide the bases for a transformational justice and a new society in-process. Holland, too, in his recent *Nomad Citizenship* (2011), has provided exemplary analysis of new societal forms as well as strategies by which they may be engendered.[11]

11. We have also been inspired by the works of Hardt (1993), Roberto Unger (2004), and Dyer-Witherford (1999).

Holland reviews many incipient as well as well-established alternative forms of social organization and applies an "affirmative nomadology," a refinement of Deleuze and Guattari's (1987) call for a "war-machine" that challenges ossified structures and their constituting dynamics.

Chapter 3 of Holland's *Nomad Citizenship* canvasses various initiatives around the world that instantiate non-hierarchical, bottom up, participatory forms of organizations. These range from the Orpheus Chamber Orchestra which is "leaderless" (no permanent conductor) compared to the model of traditional orchestras to: Mary Parker Follet's (1949; see also Graham, 2003) call for non-hierarchical styles of management as opposed to Taylorism; an introduction to Jane Jacob's (1961) concept of "public acquaintances," arguing that a new form of social interaction in the cities puts people in constant interactions with diverse others and are everyday occasions for understanding differences and for self-organization; comparing Carl Schmidt's (2007) notion of a centralized state as a protective role to that of Martin Buber's (1972) call for the State organized around the centrality of mediation to regulate the various conflicts that emerge (Holland, 40–41), a contrast that can also be situated in Schmidt valuing the Fatherland, Follet, the Motherland (ibid., 82); comparing non-hierarchical, participatory, ground-up, self-organizing internet developments such as Wikipedia, Kuro5shin (89), Slashdot, Linux, and others; global development such as worker-cooperative—e.g., El Ceibo, a cocoa growing worker organization in Bolivia, and we would add Occupy Wall Street and the Zapatistas movement, and many more.

In Chapter 4, Holland moves toward advocating "free-market communism" as an alternative to hierarchically organized political economies such as those based on capital accumulation. Following Deleuze and Guattari (1987), a socius based on capitalism as a market place must be distinguished from capital accumulation and control. Capitalism itself concerns ongoing deterritorialization and decoding, while capital accumulation concerns an abstract machine of capture, axiomatization, and hierarchy (Deleuze and Guattari, 1987: 246, 360, 386; see also Holland, 1999: 78–87; 2011: 129–135; Patton, 2000: 96–97, 99, 122–124).[12] He suggests small-scale cooperatives, that both self-

12. Foucault's work on biopower, going beyond his earlier analysis of disciplinary mechanisms (1977) thus can be connected with this axiomatization (Hollands, 2011: 132; Foucault, 2003: 242–263; 2000). Foucault's biopower and biopolitics, unlike disciplinary mechanisms that coordinate bodily forces toward some economic and political end assuring normalization, focus more on populations, "which tries to control the series of random events that can occur in a living mass, a technology which tries to predict the probability of those events.... which aims to establish a sort of homeostasis" (Foucault, 2003: 249). These are, in short, "technologies of regulation" compared to the earlier "technologies of discipline."

organize and are driven by horizontal work relations. These undermine axiomatization and systematic capture in the socius (ibid., 129–135) and release lines of flight, deterritorialization, a new flow, a more diverse flux toward a "new earth," a people yet to come (Deleuze and Guattari, 1987: 510; Patton, 2000: 107–108). In this view an affirmative nomadology goes hand in hand with deterritorialization, recoding in less static boundaries, transformative practices and "permanent revolution" or "permanent insurrection." Molecular assemblages would be modal types in the emergent socius that are always mutating and in process of metamorphosis while providing relative degrees of stability. Thus strategies must be devised "to free from capitalist command the potential of markets to self-organize social production immanently" (Holland, 2011: 104).

The central strategy, Holland suggests, is not to confront capital directly through violence, but heading Walter Benjamin's (1986) "Critique of Violence," to engender a "slow-motion general strike," a gradual disengagement with dominant, hierarchically organized molar assemblages. The call is for seeking out the many molecular forms of assemblages that have already self-organized, or are emergent and affirm an alternative logic, an affirmative nomadology that valorizes horizontal, participatory, self-organizing, experimental, and open forms of organization (see also Unger, 2004). Not confrontation against the enormous central powers of the State, but a policy of "walking away," and as Holland (ibid., 97–98) insists, "walking to" an alternative, no better expressed than by Deleuze and Guattari (1987: 161; cited in Holland, 2011: 97): "Lodge yourself on a stratum, experiment with the opportunities it offers, find an advantageous place on it, find potential movements of deterritorialization, ... experience them, produce flow conjunctions here and there, try out continuums of intensity segment by segment, have a small plot of land at all times." Metaphorically, the "plot of land" is an alternative space, be it organizational or conceptual.

We return to Bradley's model for further guidance as to how spectra and logons are constructed and potentially deconstructed. From Figure 4, the intersection of high control with high flux, which is engendered by command forms of capital logic, indicates, consistent with Karl Marx, increasing tension and contradiction as continuously released creative energies seeking expression are continuously captured within an axiomatic system. Holland's affirmative nomadology could be seen as the viable alternative, a slow motion general strike, a walking away and disinvestment in dominant forms, to outright revolution that confronts State power. Here the call is for greater flux, "permanent insurrection," and the desirability of molecular assemblages. The directions is clear, the "plot of land" and the "people yet to come" are self-organizing emergents from the attractor basins formed at the intersections of low control and high flux. This is the space in which alternative information construction could take place, a "refrain" or melody, that resonates with the many. Spectra, or holographically encoded waveform information develop in these regions, tending toward relative stability, but yet always open and sensitive to perturbation. Within this milieu a more molecular quantum holographic lexicon emerges in which word associations would tend toward basins of attractions reflecting the new non-hierarchical relations of bodies to bodies where "I" is Other, Other is "I," an ongoing becoming-other. These are holographically encoded and stored in the holofield and begin to take on greater value as they are accessed by the multitude of persons, and tend toward harmonic resonance through phase conjugation dynamics. This establishes the nuanced normative order and the particulars for the community and abstract generalized other. Here, too, differences are efficacious due to the inherent nuances in matrices instantiated (wave function collapse) reflecting the dynamics of schema QD, and indeterminacy in the everyday construction of logons, perceptions images, and diffusely instantiated "I's." These lend themselves to the further development of molecular forms of assemblages. As dynamic systems theory tells us, even small inputs will affect the emergent "dissipative structures" that will be the new form of molecular assemblages in the emerging socius ("new earth"). Agents within this space access the newly emergent spectra via phase conjugation dynamics, not linearly, but as Gabor's (1946) notion of logons suggests, always somewhat indeterminately, and thus opening space for choice and will. "I" is never the I-as-the-same, but an I-always-in-process, a mutating, becoming-other. Schema QD informs us that the immediate Other will not be constituted on the basis of distancing, reduction and diminishment, but by an ontology of the infinite Other as conceptualized by Emmanuel Levinas (1969, 1998; for an accessible introduction, see Critchley, 1999) and in the form of "I-thou" as expressed in Martin Buber (1971). Here the excessive investor in harm will give way to de-investment in the tendency to exclude, distance, categorize, subordinate, and diminish.

We can now also see how schema QD, instantiated and diagramed in matrix form with a distinct signature wave in historical political economic basins of attractor states can witness alternative forms of expression. The immediate Other will begin to take form of I-as-an-Other, Other-as-an-I, a reciprocity that is reflected in "I-thou" dialogical encounters. The community generalized other will reflect new molecular assemblage forms, a basis not of "control," but a sense of solidarity and inseparability between self, others, and community. The abstract generalized other in the form of the oedipal-based Fatherland will wither away. The ego-ideal, I(O), will as Foucault[13] informs us, take on a new form in historical conditions, one in which it is reconnected to reciprocal forms of othering. The speaking subject, "I," will find a plurality of discursive subject-positions[14] in which it can take up residence in narrative constructions. The cut of the subject, or the collapse of the wave function representing the plurality of possibilities of expression of the person, will represent not static identities, perception-images, and distinct ever enduring realities, but an acknowledgement and celebration of differences and the often contingent nature of these constructions. Becoming-other, becoming-minortarian will be a modal type in the emergent socius. The signature waves of interacting persons would form interference patterns and each would leave a permanent trace of her/himself in the Other: "I" becomes Other, Other becomes "I."

3. Becoming Identities

A transformative justice would follow not only a change in current political economy toward molecular forms of assemblages, with its attendant form of expression ("minor-literature"), but will witness a necessary reconstruction of traditional identities in our current criminal justice system. Restorative justice has begun the metamorphosis, but it is toward a transformative justice we must look for more genuine social justice. The work of Pavlich (2005) and Woolford (2009) have provided much illuminating analysis as to how traditional criminal justice identities can be deconstructed and reconstructed. The identities of victims, offenders, communities, and facilitators should be reconsidered along the lines of contextual, contingent, relative, ironic, and dialectic developments, but mindful of the always-present danger of usurpation. Consider, for example, those who have been victimized by harms of repression and reduction and their narrative constructions of injury. "Trauma narratives," as Woolford (p. 95-98) argues, differentially resonate[15] based on connectedness with dominant understandings. Some, then, are more resonating, striking a cord (no pun intended), some are not. Disenfranchised group's trauma

---

13. See Foucault's (1982, 1994) essay on "technologies of the self."

14. Unger (2004) for example recommends "role jumbling" in which one experiences various roles. This necessarily would implicate being introduced to various quantum holographic lexicons and their associative words.

15. His use of "resonate" we use literally rather than figuratively. As we have seen, it is connected with phase conjugation, harmonic resonance, opening channels of information flow.

narratives are the subject, however, of reformulation by powerful groups in sustaining their vision of reality. Thus "resonance" initially established between victims' and offenders' narratives and others could be transformed by conservative, system-sustaining forces that reformulate the nature of responsibility and blame, excusing the political economic order of its contribution, such as we find in blaming-the-victim ideologies.

Trauma narratives are in need of being resituated in broader socio-political-economic contexts. A socius of a plurality of horizontally organized molecular assemblages is more open to a wide information flow. It is about the discovery of the complexities of the Other. Consider, for example, Maruna's *Making Good* (2006) where, researching recently released inmates, he shows the efficacy of a "redemption script" in desistance from crime. This is opposed to a reactive "condemnation script" which is a narrative of despair, a resignation to seeing oneself as merely a pawn devoid of agency, a fatalism that militates against desistance from crime. A redemption or generative script, on the other hand, acknowledges a bad moment in biography, but the ever possibility of change, transformation, a "new me," a self-rebiographing. Within this model of transformation, we heed Maruna's (ibid., 166) point that "cultures with few models of redemption may be the cultures with more doomed deviants." This is a call for expanding possible discursive subject positions within which an "I" can take form; it suggests that these subject positions have been greatly restricted for the returning ex-offender who has paid for his crime. Our emergent transformational form of justice would foster redemption scripts and "replacement discourses" (Henry and Milovanovic, 1996) providing greater opportunities of erring individuals to reintegrate into society.[16] The collapse of the wave function could take on more diverse forms portraying the complexities of the human condition. Pacifying, disciplining and channeling offender's narratives and identities into molar forms as in our current criminal justice system are ever present dangers. They tend to produce encoded holographic information that when accessed provides a narrow vision of causation, responsibility, will, choice, and opportunities.

More recent suggestions in trauma therapy approaches, moving beyond a classical-based model, have suggested a shift to a quantum or neuroquantology based theory of psychiatry (Mender, 2010; Globus, 2010) as well as to a "holographic reprocessing" model (Katz, 2005). In the latter model traumatic memory experiences of self and others are said to have been encoded holo-

16. Maruna (2006: 167) notes that "not only are there few change stories readily available to ex-offenders ... there may not even be a language or discourse available for describing this change. The words may simply not be there."

graphically and continuously reappear in a person's life until holographic reprocessing therapy revisits the traumatic event to recode it in less traumatic resolutions. Some empirical studies of the latter have shown significant results (Basharpoor *et al.*, 2011; Katz *et al.*, 2008, 2013). As of the present day, traditional therapeutic approaches seem reluctant to entertain this new framework. The current challenge is to incorporate both quantum *and* holography theory in devising a new paradigm. This will be especially necessary in the recent developments and application of "virtual reality exposure therapy" in treating post-traumatic stress disorder (PTSD) as well as other issues by virtual reality reconstructions, an interest being followed keenly by the military for veterans of Vietnam, Iraq, and Afghanistan wars. Currently, even though this holographically suggestive model is showing some sign of success in practice, the underlying theoretical framework still remains predominantly in the classical-materialist paradigm. It would seem that this new thrust would greatly benefit by looking at concepts such as phase conjugation, creation of new spectra, nonlocality, tunneling, collapse of the wave function, as well as engaging some framework of agency. Further engagement with Seymour Epstein's (2014) "cognitive-experiential self-theory,"[17] often cited in holographic reprocessing literature (e.g., Katz, 2005), could lead to a productive integration with many of the authors cited in this work particularly in emerging virtual reality exposure therapy (VRET).[18]

The notion of "communities," Woolford (2009) poignantly argues, is in need of reconstruction. Citing Pavlich (2001), even progressive notions of community tend toward exclusionary practices, physically and/or symbolically. In place of static, closed notions an alternative concept of "becoming communities" (Agamban, 1993; Fraser, 1997, 2000; Woolford, 2009: 107–109) has been suggested which is more porous, flexible, open to the Other, and is only understood by members in their felt experiences and in their constitutive constructions. This would be more along the lines of molecular assemblages. Pavlich's concept of "hospitality" (2001, 2007; see also Woolford's discussion, 2009: 143–145) is provided as an alternative to traditional notions of "community,"

17. Epstein (2014) argues that mind is dualistic: one part is the rational-analytic mind, the other, intuitive-experimental mind which would include the emotions, affect. This formulation has compatibilities with the works of Marcer and Schempp and others in holographic theories of the mind. However, an exciting challenge will be to integrate Epstein's work more closely with both quantum and holography theory. Certainly Katz (2005) makes an initial foray into this necessary integrative work in her holographic reprocessing therapeutic approach.

18. See for example: Hoffman (2004), Nelson (2012, 2013), Wilson (2013), Sippel (2013), "About Virtual Reality" (n.d.)

where the "good host" is one who actively invests energy to know the Other, and the Other to know him/her, especially where distinct differences exist. This enhances a form of othering in which "I" is becoming different, becoming-other. This provides a different basis for the Other and for the ego-ideal, I(O), in Schema QD, leading to an alternative collapse of the wavefunction, instantiating a non-molar understanding in discursive expression and for the activation of a more molecular discursive subject-position within which an "I," as speaking subject, can be situated in constructions of social reality.

Another role in need of deconstruction is that of a facilitator, mediator, or negotiator in dispute settlement. The identity of "facilitator," Woolford tells us, often finds overly professionalized expression in restorative justice programs, creating distance between the actual and its representations. Alternative indigenous models have been often cited in the literature. Consider in this context the use of previous gang members as "interrupters" in Cure Violence programs taking hold in larger cities in the U.S. Interrupters are strategically situated in collapsing the wave function otherwise to traditional ways leading to the creation of stories that justify or rationalize retaliation and violence.

Kuttner (2011), has also provided a useful starting point, a quantized analysis[19] of dispute resolution arguing that two orientations exist, derived from an understanding of the inherent duality of the wave function as particle *and* wave. The dominant form of dispute resolution privileges the particle-like rather than the wave-like nature of the wavefunction. The particle-like orientation in negotiations assumes: individuals pursuing their self-interests; a prioritization of categorizing interests; a focus on clarifying each party's issue; and resolution via linear rational logic. It is more like Buber's (1972) I-it dynamic. A wave-like orientation, on the other hand, assumes a more dialogical co-construction of interests and possible solutions, closer to Buber's I-thou relation. The wave-like orientation, Kuttner (ibid., 347) tells us, "allows the participants to establish an active self-organizing system in which to explore new ideas." It does not always follow linear logic. Here, a "suspension of foundational assumptions, rather than their defense, is welcomed," and "preconceived knowledge of the problem that needs to be solved and how it should be solved is replaced by an ongoing mutual discovery of the problem parties are facing" (ibid.). Discovering the problem without preconceived formulations, then, paves the way for the problem to emerge in dialogue (ibid.). Kuttner, in anticipation of a quantum holographic approach, does use the language of "tun-

19. Kuttner, unfortunately sees his usage of quantum theory by analogy where we view it as isomorphism.

ing" in each other's willingness to engage in wave-like negotiations. We again use this literally not figuratively or metaphorically: the former suggests phase conjugation dynamics, establishment of standing waves that are in harmonic resolution, information creation and flow. This can be further developed from Wheeler and Morris (cited in Kuttner, p. 354) who recommend that respective parties "be attuned to the other party's verbal and non-verbal cues," and to focus on "pitch of voice, speed of conversation, pauses, and verbal stumbles" for more holistic understandings. Kuttner does conclude, however, that most successful negotiations entail both orientations (i.e., both wave-like and particle-like). In the current broader context of political economy this would probably be the most pragmatic; however, the wave-like orientation would be an on-going emergent in a socius characterized by ubiquitous molecular assemblages.

Kuttner's approach could be better situated in both quantum and holography theory. We could, for example, begin to look at phase conjugation dynamics underlying information production in negotiations; how a common conceptual space is created along with the quantum lexicon that provides associative linkages of words that are more "thou"- than "it"-oriented; how access to this space provides the basis for the construction of perception-images and a speaking-being, an "I" that is the speaking subject. In this context, we would examine: the independent reality forms of construction of the negotiators; the potential for a rethinking of more dominantly accepted forms and what this might mean in subsequent reconstruction of memory-images; how these forms of wave-like negotiations could contribute to an ongoing development of a replacement discourse, a minor-discourse, a becoming-minor; and how, beyond the immediate negotiations, a substantial contribution to the holofield of holographically encoded information takes place, the access to which becomes a resonating point for further transformative discourse; etc. Wave-like negotiations, then, could contribute to the development of replacement discourses that undermine static particle-like axiomatic systems.

Returning to Holland, a slow-motion general strike sets the grounds for the withering away of traditional reifying axiomatic systems. A new discourse reflecting a respect for and valorizing of far-from-equilibrium conditions and self-organizing emerges, not bound to a rigid axiomatization, but a more discourse-in-process, a becoming-minor of discourse. Holland's affirmative nomadology focuses more on the "content" dimension of an assemblage. Looking at the other end of this dimension, "expression," we can see how a new regime of signs ("expression") could emerge, reflective of new forms of interactions ("content") amongst human beings-becoming. This would be in accord with Deleuze and Guattari's call for minor-literature. We would begin to see a greater populating of the quantum holographic lexicon, a wordspace within the holofield of hologram word-forms representing both community commonalities and individual differences, accessible by community member's signature waves, by way of a the mapping-to and mapping-from phase conjugate dynamics suggested by Marcer and Schempp (1998, 1999). This lexicon storehouse, residing not in the brains of individuals but consistent with Bergson, Laszlo, Mitchell and the notion of nonlocality, in an external holofield, would tend not to be populated by order words, rigid noun forms, and master signifiers that resist change, nor with the organic regime form of signs that privileges linearity, homeostasis, and images reflecting static cuts of blocs of spacetime out of the plane of immanence; rather, with the emergence of ubiquitous molecular assemblages, forms of expression would be more along the lines of a crystalline regime of signs, an abundance of verb forms, master signifiers that are more amenable to change, and mobile cuts of spacetime.[20] At the same time, consistent with assemblage theory, territorialization will also appear with an active deterritorialization dimension as that which is created is ever open to be rethought and reworked. Undoubtedly, as Roberto Unger (2004) has argued in his "empowered democracy," moments of differences and tensions would increase, but would not be the basis of distancing and reducing the Other, but moments of a deeper understanding about the complexities of being human. Affirmative nomadology is more in line with Deleuze's "crystalline regime of signs," whereby greater zones of indetermination are created within which novel responses can emerge, where an ongoing falsifying discourse sustains permanent change, where a new ethics will materialize. To these we will turn in the next section.

In sum, models for the potential transformed society, a permanent revolution, or better "permanent insurrection" (Holland, 2011: 97) already exist. A "slow-motion general strike" is already upon us. But capture and axiomatization await the new movements, ready to imprison them within molar assemblages, dominant discourse (major literature), and restrictive discursive subject-positions. Quantum holography indicates, with the inspirational works of Deleuze and Guattari and their interpreters, that a new socius is possible, one in which active recognition, novelty, mutation, and metamorphosis of self, others and society can prevail.

20. This is the thrust, as we understand it, of Bohm's call for a new language based on process, change, transformation, "a mode in which movement is to be taken as primary in our thinking and in which this notion will be incorporated into the language structure by allowing the verb rather than the noun to play a primary role" (1980: 30).

Discursive Reconstructions: "Man of Truth" v. "Power of the False"

Deleuze has provided some analysis of how truth is constructed in specific forms of regimes of signs. For example, consider Deleuze's (1986) discussion, following Nietzsche, of the "man of truth." In his two-volume cinema books he contrasts two orientations of semiotics: an "organic regime" rooted in more linear constructions of truth based more on a Hegelian and Lacanian notion of desire as overcoming lack, where truth is discoverable, is knowable awaiting recognition, and which supports an ontology of being; as differing from a "crystalline regime" where the image can always be instantiated otherwise, in our view, due to particular forms of wave function collapse, and where truth following Nietzsche is created, an artistic act, part of an ontology of the "power of the false" and becoming. The organic regime spawns a system of judgment and the "man of truth" (Deleuze, 1989: 140–142; Rodowick, 1997: 85, 134–138; Bogue, 2003a: 147–150).[21] This person claims access to a higher truth, who judges life according to these transcendental, universal values, and who often acts to revenge perceived wrongful acts. Elsewhere (Milovanovic and Henry, 2001) we have referred to this person as an excessive investor in harms of reduction and repression. His is a phase conjugate resonance with a lexicon imbued with notions of a rigidly ordered society, where deviations are signs of chaos and disorder. It is also a form of othering that excludes, objectifies, diminishes and distances (Young, 2011: 64). It is the breeding ground for the excessive investor in harms of reduction and repression. This is Nietzsche's exemplary manifestation of nihilism, a reactive mode, and a form of the will to power reflecting a "base" form, a will to dominate. It is also, we argue, the underlying ontological framework of our criminal justice system, the support of which rests with its continuous success in its failure. Consider the zero-sum nature of the trial, where one truth is said to ultimately prevail as the Truth. It can be discovered, described and rationally explained.

Alternatively, the crystalline regime supports the "power of the false," the possibility of always being otherwise, of change, mutation, becoming. Here truth is created in process, a part of becoming: the good is evaluated as the inherent power of the false, a will to power not as domination but growth, becoming, transformation. As Rodowick (1997: 85) summarizes, it is about "how

21. Consider Rodowick (1997: 85): "In the organic regime, the form of truth requires the false as is its opposite or negation in order to master the negative under the form of unity and identity."

thought can be kept moving, not toward a predetermined end, but toward the new and unforeseen in terms of what Bergson calls the open or 'creative evolution.'" This is the "artist, a joyous forger or falsifier who creates value by affirming the becoming of the world" (Bogue, 2003a: 149). It is also Rorty's (1989) concept of the "good prophet" who transcends the dictates of common logic. Thus the two regimes of signs differ: "for the former [organic regime] it is the *discovery* of concepts through negation, repetition, and identity toward ever more self-identical Being; for the latter [crystalline regime] it is the *creation* of concepts through difference and nonidentity in a continually open becoming" (ibid., emphasis in original).

The organic regime of signs is based on an ontological foundation of movement-images. These images situate action in mean-ends circuits; they reflect assumptions of the independence of objects, and an objective reality preexisting intervention (thought) practices (Rodowick, 1999: 89; see also Bogue, 2003a). Desire, in this model is based on lack and necessity for closure. Here, Hegel is inspirational for the classic works by Russian film director Sergei Eisenstein and Lacanian-based cinema analysis. The crystalline-regime of signs, alternatively based on an ontology of the time-image "is provisional and contingent. It replaces the object continually 'erasing' it and creating it anew, giving way to other equally adequate descriptions, which may modify or contradict the ones preceding it ... description loosens the anchor of designation that attaches images to things" (Rodowick, 1999: 90). This latter approach is rooted more in Nietzsche and Bergson, according to Deleuze (1986, 1989). Film directors such as Alain Resnais, Orson Welles, and Alain Robbe-Grillet exemplify application of this form to cinema. It encourages not a linear development of narratives (and man of truth) based on a presumed, pre-existing objective reality, reconstituted in inattentive recognition/judgment practices, but by way of attentive recognition/judgment where truth can always be otherwise, is provisional, contingent, and nonlinear. The power of the false is a destabilizing orientation, encouraging deterritorialization, experimentation, and creation. It actively engages a zone of indeterminacy where, via the circuit diagram dynamics, various versions of reality can exist side by side, a quantum superposition, until collapse into one that becomes the virtual image that is instantiated as a basis for action. Deleuze (1989) finds in Jean Rouch's ethnofictions (see his "Jaguar", "Moi, Un Noir," "Les Maitres Fous") the basis of a "participatory ethnography" that dialogically creates reality. It is meant to be provocative and self-inventive (Bogue, 2003a: 153–154).

In our quantum holographic adaptation, we can see how the organic regime of signs tends to create images and reflective signs (semiotics) from movement in the plane of immanence, the basis of which can be captured by a linguistic system organized by the orthogonal axes of paradigm and syntagm where the signifier is said to stand for a signified. Resonance is established between everyday linguistic usage and the underlying images derived and reflective of the plane of immanence. A "cut" of a "bloc of spacetime" in the underlying flow produces static reality constructions that can be categorized and brought within the purview of practices of normalization. An agent, finding her/himself in this milieu, is provided a constellation of signifiers, or associative word linkages, that provide the basis for reality construction. This is the mapping to and from (signifier, signified), a phase conjugation suggested by Marcer and Schempp (1998, 1999). Phase conjugation dynamics produce both continuity with this underlying image flow and reification.

Logons constructed in this circumscribed quantum holographic organic lexicon tend toward dominant forms, although indeterminacy assures incongruencies, a space where things could be otherwise. The crystalline regime of signs, on the other hand, always suspends any possible creation of a linearly constructed Truth. Images reflecting movement of energy in the plane of immanence are also in flux, defying clear cut distinctions. A dominant linguistics still attempts to capture these images in molar forms; however, emerging constructions, or minor-literature, always insists, waiting idealized expression by cultural revolutionaries. Here, the "cut" is always a "mobile cut of spacetime," resisting precise categorization and normalization. The agent who is the interlocutor can only be a subject-in-process, a becoming-other. It is here where a premium exists on experimenting, suspending beliefs, negotiating boundaries, and embracing what Foucault in his late works called "limit experiences" by not only approaching but crossing boundaries (Lyng, 2005: 41–47) and experiencing what a body can do, what it is capable of, its power to effect and be affected. We conceptualize this quantum holographically as moments in which mobile cuts of spacetime produce novel information, which can accrue in a milieu, perhaps becoming a refrain, a tune, a basis of the development of a molecular assemblage.

It is here where cultural criminology is providing key insights as to social constructions by the media, in their explanations of various loops and spirals, spinning, in our view, reality in organic linear forms, assuming as truth the otherwise imposed ontology of lack and manipulated desire. Reality TV programs such as ride-alongs in cop shows provide a point of subjectification, a discursive subject position within which a viewing "I" can take up residence in understanding what is going on. In other language, cultural criminology's critique of media practices concerns their complacency with the dynamics of an otherwise induced lack and the continued commitment to the Oedipus structuration. As Ferrell *et al.* (2008: 154) argue, "when crime policy is made in the media, when courts echo with media-made expectation, when police officers perform for their own cameras, criminologists must find ways to penetrate these dynamics if they're to humanize them. When crime collapses into commodity, war into entertainment, reality into virtuality, criminologists must find new forms of intellectual activism appropriate to these confounded circumstances." A reorientation to the crystalline regime and molecular forms of assemblages provides an alternative basis of activism, a path, not necessarily an end, in which new constructions of social reality can begin. Here "cuts" of mobile blocs of spacetime would bring out the fluidity and contextuality of social constructions. It would embrace active attentive/judgment and the potentials of change, mutation, and metamorphosis. A new relation between "control" and "flux" would produce alternative logons, information reflecting the complexities of being human in interaction with others. This is indeed "dangerous knowledge" (ibid., 158), which asks us to put in suspension organic regimes of signs connected with molar assemblages of late capitalism, and to embrace an alternative ontology. This resituating creates alternative images of being-otherwise, always, however, subject to the potentialities of capture into yet new forms of axiomatization. A "liquid ethnology" (ibid., 182–184) is one strategy offered compatible with the logic of a "the people yet to come" that is "immersed in the ongoing interplay of images ... [an] ethnography comfortable with the shifting boundaries between research, research subjects, and cultural activism." They (ibid., 184–188) also suggest a "visual criminology" and provide some directions.

However, cultural criminology's position on visual criminology is in need of a new ontological basis. Deleuze's two-volume cinema books provide a valuable alternative with the discussion of the crystalline regime of signs. His resituated ontology of image construction and how it can be otherwise is suggestive of alternative constructions: i.e., the power of the false that destabilizes ongoing constructions of reality; falsifying narratives that lead to reconceptualizing reality in more dynamic discourse reflecting the interconnected nature of reality; generation of points of indiscernibility in otherwise linear constructions; providing "seed-crystals" that much like the mythical insurgent Trojan horse lead to alternative actualizations that destabilize hierarchical stabilities; the creation of lines of flight that are not reifying but generators of new images; crystalline-based ontologies of film constructions such as in "ethnofictions," "participatory ethnographies," and "fabulation" (e.g, alternative story-telling and identity construction) suggested in the films of Jean Rouch that provoke and invent new stories of being and becoming; "cinemas of the body" that bring out the yearnings of the repressed and marginalized; in short, reconceptualizing the linearities of the organic-regime of signs—mimicked in contemporary media, police, legal, scientific practices, and theorizing in criminology, forms of axiomatization that make ontological assumptions rooted in the classic-materialist paradigm of Newton, generating theorems and objective Truths (Bogue, 2003a: 180)—toward a problem approach where reality could always be otherwise, where many contingent truths exist, where a choice and valuation must be made, where not stasis but permanent becoming, mutation and metamorphosis are ontologically rooted in the process-informational paradigm. But this is not the pursuit of establishing a system of judgment, an organic regime's ultimate call for the man of truth who judges from some transcendental perspective, what Nietzsche saw as the manifestation of reactive forces, but a pursuit of the "will to falsehood" (Deleuze, 193: 184; Rodowick, 1999: 134–135), artistic expressions that do not attempt to discover truths, a pursuit of creating truths in a world of often inexplicables, undecidabiles, and incompossibilities (Rodowick, 136). Truth is arrived at by an ongoing "series of falsifications" of dominant truths. As Rodowick (ibid., 137) says, "falsifying narration augment our powers of life by affirming change and by creating images of thought that put us in direct contact with change and becoming as fundamental forces." Cultural criminology, in short, can benefit from this novel approach in truth constructions in arriving at strategies that tend toward disruptions and becoming-otherwise.

Quantum holography can be integrated with Deleuze's cinema books and provide cultural criminology with a new basis of information construction and transmission. Information construction and transmission are provided a physicality and specification as to how, on the one hand, encoding, harmonic resonance, understanding, evaluations, and solidarity take place, and on the other hand, how evaluations and matching of resonances can also lead to incongruences in resonance and hence the basis for new evaluations, which directly implicate schema QD and inherent indeterminacy in social reality construction processes. The wave function, we have seen, represents clouds of possibilities. It is this wave function that is subject to capture and axiomatization in organic regimes of signs. And it is this axiomatic system that shapes the possible forms a community and abstract generalized other can take. But the wave function can be, we have also seen, collapsed otherwise bringing out novel instantiations, new imagery, new signs, new signifiers that are more process oriented, contingent, relative, and always mutative. Emergent expressive dissipative structures would include an increasing influx of a quantum holographic lexicon, semantic spaces reflecting subcultural variety, encoded holographically in the holofield and accessible by resonating signature waves of community members, but ever in variation and indeterminate in the everyday constructions of logons.

We see elements of the crystalline regime and the power of the false in some forms of restorative justice, but it is toward a transformative justice we must look in order to find its ontologically more natural state. In late modernity, with its insecurities, exclusionary practices, commodification of work and leisure, the man of truth is a dominant response. A system of judgment and the man of truth are embedded as a logic in the axiomatic system. Here the quantum holographic lexicon, including concentrations of master signifiers (Lacan, 1977), or "shining points" (Bergson, 2002), accent the world in terms of stases, determinacy, inevitability, closure, and inherent desirability of linear logical thought. Affordances carried along by lines of flight (wave function), much like Bohm's soliton-acting "pilot waves," provide a narrow range of possible instantiations. The abstract machine assures receptive agents (inattentive recognition/judgment) who, advertently and inadvertently, constitutively reconstruct and reify dominant molar background elements within a subterranean conceptual space (Wendt's "social wave function") encoded as holograms in the holofield (ZPF). Contemporary identity politics and a pursuit of dogmatic forms of rights are but derivatives and reifiers of this axiomatic system.

Alternatively, Deleuze advocates a "people yet to come," identities that are becoming, in-process, mutating, and transformative, and a more pragmatic view of rights that is more like case law with its inherent logic of adaptation and change. The latter he refers to as "jurisprudence" rather than a system of rights. It is not about enduring universal, abstract, natural rights attached to isolated persons, but is about invention, creation—"it operates within a social field to resolve specific problems and create rights [ongoingly]" (Lefebvre, 2008: 55). Hence, in Lefebvre's (ibid., 58–59) reading of Deleuze and Bergson, it is an orientation that privileges problems as encounters demanding attentive judgment/recognition, and, as we have developed, privileging the power of the false and the creation of ever-new concepts that reflect the evolving flux that is human interaction.

Evaluative Schemas: Distributive and Retributive Guideposts

This provides a segue to one aspect of a transformative justice, which is the issue of ethics. Previously we (Milovanovic, 2011) have mapped several justice-rendering schemas, systems of justice that theoretically could be stand-alones. We specifically addressed the issue of a transformative justice model (ibid., 26–41). The concepts of deterritorialization and reterritorialization implicate an evaluative schema by which a more genuine justice can be developed. We are now in a better position to argue that quantum holography offers us the key concepts needed to be integrated with this work in giving impetus to a process-informational paradigm. We will focus on two principles (distributive, retributive) that are in accord with integrating quantum holography into the previous analysis of transformative justice. We are mindful of Woolford's (2009: 150) caution against developing blueprints and, agreeing with him, see our exercise more as providing some guideposts.

Activist thought is often criticized for its reticence in developing normative theory and concrete criteria that may lay at the basis of ethical standards (Litowitz, 1997). We draw from Patton's (2000: 59–67) reading of Deleuze's (1983) *Nietzsche and Philosophy* and Hardt's (1993) *Deleuze*, a work on Spinoza in developing an alternative notion of desire and an evaluative schema by which it could be measured. Our main critical target is the Hegelian-inspired work of Jacques Lacan on the nature of desire as purely a response to lack situated in an Oedipal structuration.[22]

Nietzsche suggests that a "will to power" and active and reactive forces are the ontological basis of human activity. Will to power must be read as a desire to increase one's ability to effect and be affected. Reactive forces are those that impose repetition, stasis, reconfirmation of order and the same; they are "forces of limitation ... which resist the activity of other forces" (Patton, 2000: 61). Active forces are those that seek to push to the limit of the possible and to mutation, change, and transformation (Patton, ibid.). It is genealogical analysis which will suggest the quality of each at play in a spacetime circumstance (ibid., 62). This is averse to "any simple axiological priority of active over reactive, affirmative over negative conclusion" (ibid., 63). Evaluation, then, must consider genealogy, an influential concept developed by Nietzsche (2003) in *The Genealogy of Morals*, and refined in Foucault's (1971) essay, "Nietzsche, Genealogy, History." It is the science of looking at historical development as more nonlinear, contingent, emergent, interactive, ironic, relative, and dialectical.[23] Nevertheless, the will to power is not a primarily negative force (ibid., 64).

---

22. Deleuze and Guattari (1983), in *Anti-Oedipus*, explained how the Oedipus complex and desire based on lack found a supportive role in capitalism. They, alternatively, build a model of desire based not on an inherent lack, but on production and becoming. It could be argued, however, the vestiges of the Oedipus in close alignment with certain forms of capitalism remain ubiquitous, and with effect.

23. For a particular clear explanation, see May (1994: 98–119). Genealogical analysis disavows the pursuit of discovering essences, it (ibid., 90–91) "does not look for a unitary origin, a single source from which its object springs ... it is a historical account of its object, one that holds history to be contingent, dispersed, shifting, and without a goal ..." Referencing the work of Deleuze and Guattari, May (ibid., 108) says, "they offer an outline of how a genealogist might go about discussing the state, as abstract machine rather than institutions, instantiated not only at the macropolitical but also at the micropolitical level, reliant

Evaluation would make a determination of the existence of more molar forms of assemblages or more molecular in place, the latter always experiencing the disruptive, regenerating effects of deterritorialization. The latter are the generators of "lines of flight" toward limits, and beyond, putting into variation what had been instantiated as one particular collapse of the wave function from the many possible inhering in the cloud of possibilities on the plane of immanence. Thus "good" would mean configurations of forces in a socius or assemblage that provide wide latitude for deterritorialization and lines of flights (read as affordances) toward and beyond the limits, increasing agents' capacity to effect and be affected. "Bad" would be reactive forces that close of potentialities and the will to power. This evaluation is always mindful, Patton (ibid., 65–66, 87) cautions us, of "lines of destruction" or reactive forces.

Hardt's (1993: 54) discussion of Deleuze's ethics is instructive as to alternative notions. He notes that Deleuze uses a Bergsonian approach to critique Hegel and to supplement Nietzsche in arriving at a reformulated Spinozean more socially-centered ethics. He (ibid., 37, 57) notes that Deleuze's earlier reading of Nietzsche's work falls short as to a more comprehensive picture as to real social struggles, needs a theory of practice and agency in response to forces, and is overly an individualistic theory that needs a social dimension ("sociality"). Accordingly, Deleuze turns to Spinoza for providing this additional dimension. For Spinoza, it is a contrast between "sad passions" and "joyful passions"[24] that take place inter-subjectively and intra-subjectively that must be integrated. Spinoza's view of power concerns the power to effect and to be affected. His model values affirmative joyful passions in societal interactions that will increase the power to effect and be affected, and hence, to return to Deleuze, promote a continuous becoming, an ongoing mutation, change, transformation. Hardt examines Spinoza's question of what the body is capable of. Any encounter between people (inter-subjectivity) is one where joyful or sad passions can be activated and cultivated. Thus, for Hardt (ibid., 92), "our power to be affected is filled by active affections, it relates directly to our power to act, but to the extent that it is filled by passive affections, it relates only to our power to feel or suffer. Passive affections really mark our lack of power." "Joyful encounters" increase our power to effect and be affected. It "increases our power to act" (ibid., 94). Sad encounters diminish our power to act. Thus

upon local practices that sustain it, and offering always the possibility of escape from the overcoding it attempts to impose."

24. As Hardt and Negri (2009: 379) define it, "the path of joy is constantly open to new possibilities, to expand our field of imagination, our abilities to feel and be affected, our capacities for action and passion."

Hardt (ibid., 95–96) summarizes both Nietzsche's encouragement to be active and Spinoza's to become joyful in a composite Deleuzian ethical statement: become active, become joyful, or in a more dynamic statement, an ethical task of becoming active, becoming joyful (ibid., 108). And to this is added, cultivate the spirit of asking the right question and developing "adequate ideas": discover what your body is capable of becoming. Accordingly, and to bring the discussion into a more societal, inter-subjectively-based realm, "when our mind forms an idea of the common relationship shared between this body [the Other] and our body (a common notion), the joyful affection ceases to be passive and becomes active" (ibid., 99; Hardt and Negri, 2009: 379).[25]

A "common notion" which engenders joyful passions and hence increasing the power to effect and be affected, is accommodative to the notion of harmonic resonance (phase conjugation) established in the social field. This in turn, applying quantum holography, provides the grounds for the phase conjugation dynamics, an informational channel, that afford greater resources for action. The common notion could be seen as a social wave function encoding holographic information about member's unique "patterns of physical sensations, feelings, thoughts, behaviors and relations by which they experience the group as a social constancy" (Bradley, 2011: 208). This, too, is the basis of the form that the community generalized other takes, shaping the normative subcultural order. But it remains in continuous variation; as clouds of possibilities, instantiations are provided opportunities for varied expression. This wave function, thus, has a unique signature frequency to which members can become attuned in creating resonance patterns and informational channels, and thus increasing their powers to effect and be affected. Referencing schema QD would advise us that ongoing interactions with this social signature wave are not linear in construction, but drawing from Gabor (1946), can be actively or passively negotiated in both reproducing old forms, and producing indeterminate logons which provide variability from the social wave function's classic forms (repetition) of collapse. Where a threshold level of discordance exists, or where an appropriate balance of flux and control do not exist, new evaluations produce new spectra (See Mitchell and Staretz, 2011: 14; Bradley, 1998: 131).[26] In

---

25. Hardt summarizes Spinoza's point that non common notions, or discord, are the basis not of unity, but conflict, and hence produce sad passions and diminished power to effect and be affected. This has a static nature to it, and we would rather argue that discordance, or where non-resonance exists, a productive re-evaluation may lead to a higher form of understanding that is mutually encompassing.

26. Unlike Bradley, however, the "instability" engendered by an imbalance between flux and control should not be necessarily the basis of despair, for dynamic systems theory has

other words, a shift in the interference pattern, which is the social signature wave, can take place and become a new basis of "common notions" with repetition, and a community generalized other. Recall Bradley's application of Gabor's notion of logons (1998: 131): within the social field, the intersection of flux and control "generates information as a moment-by-moment—quantized—description of the collective's internal organization, encoded in terms of both structure (spatial-temporal position) and flux (distribution of energy). The succession of descriptions within space-time and spectral coordinates are logon-like units of information that are constructed and communicated, via a holographic-like process, throughout the collective." Bradley (ibid., 132) also reminds us that the logons, being overlapping analog waves, not only have degrees of indeterminacy, but each, because of the overlap, has information overlapping with future logons thus providing an anticipatory dimension for members within a collective, which in our development here, further provides an increased power to affect consequent situations. Of course, as Bradley says, normative or cultural orders provide acceptable limits to materializing these anticipations.

Lefebvre's last chapter in *The Image of Law*, also draws from Spinoza's work on the passions. He compares evaluating "sad images of law" to joyful ones. Sad images or passions exist when judges profess to only subsume the facts of the case under some legal principle, relegating themselves to an inattentive judgment mode, reifying given legal concepts. Joyful passions describe entertaining encounters in an attentive judgment mode, in the process, initiating the dynamics for creativity in decision-making. Lefebvre provides an example: the high Canadian Supreme Court, in *Delgamuukw* (1997), was to overturn an earlier court decision which could be read as a sad image of law, in providing new ground for aboriginal or "native title," a joyful form of images of law, an increase in aboriginal power to effect and be affected. Here, the case was approached as an encounter and new concepts were developed in concluding indigenous rights to aboriginal land ("native title").[27] Thus law, which cuts up a plane of immanence in static blocs of spacetime has both the potential for limiting lines of flight and becoming ossified, or for encouraging lines of flight

indicated that far-from-equilibrium conditions witness emergence and self-organization, form of dissipative structures. Consider, for example, an ensemble of jazz players. Bradley expresses some deference to jazz players, but could extend it in the direction of dynamic systems theory as suggested, for example, by Holland's (2011) *Nomad Citizenship*.

27. In 1992 the High Court of Australia, in *Mabo v. Queensland (No 2)*, ruled in favor of "native title" and against the principle of *terra nullius*, a principle used by colonizers to claim land because it was "empty" or uncivilized.

which push the possibilities to the limit, and beyond into new law (see also Unger, 1996).

From this preliminary discussion, we draw two ethical principles[28] as guidepost for thought, one, a distributive, the other, a retributive justice. They could be viewed as potential information encoded on boundary surfaces of a socius, providing lines of flight or affordance, reverberating solitons traversing the fields, that are the basis of active schema QD negotiations producing new ongoing forms of "mobile blocs of spacetime." Generated logons would not reconstitute conventional understandings, but are always somewhat indeterminate, as Gabor has spelled out. This is in accordance with Deleuze and Guattari's call for the new revolutionary agent, ever new reconstitutions of identity, a "becoming-minortarian," and with Holland's (2011) call, for an affirmative nomadology.

Distributive Justice:[29]

> "'From each according to [her]/his abilities, to each, according to [her]/his needs' and desires; to and from each the active promotion of the power to effect and power to be affected; tempered, as much as possible, by the promotion of differences and antirepresentationalism, and subject to genealogical evaluation of forces."

Retributive Justice:[30]

> "A retributive justice evaluates: degree to which measures deliver a transparent message of disapproval in a least-restrictive, collaborative, transparent, and transformative form; active and meaningful participation of those directly affected by harm, with involvement, as much as possible, by those indirectly affected; opportunities for on-

---

28. For initial thoughts, see Milovanovic (2011).

29. This is modified version of the distributive principle found in Milovanovic (2011).

30. This a modified version of the retributive principle found in Milovanovic (2011). Our principle is a synthesis from a number of contributing authors, not all of whom would necessarily self-identify with a transformative justice nor with poststructuralist/postmodern thought. Key contributing works have been by Pavlich (2005), Acorn (2004), Woolford (2009), Woolford and Ratner (2008), Arrigo and Milovanovic (2009), Milovanovic and Henry (1991, 2001), Maruna (2001), Braithwaite (2002), Parker (1999), Cornell (1998), Sullivan and Tift (2005), amongst the most prominent. Our synthesis, however, is driven by the works of Deleuze and Guattari. Along similar lines, a "virtue-based ethics" (Bersot and Arrigo, 2010) and an ethics of "dignity" (Ward, 2010; Beyleveld and Brownsword, 2001), in our view, could be included as a partial operationalization of our postmodern retributive justice.

going reconceptualization of victim, offenders, community, networks and other social assemblages, and their implementation into practice; where employment of incarceration, prioritizing respect, concern, and redemption scripts and practices; diminuation of subjugation and domination in institutions and social practices (assemblages), and privileging active, molecular lines of flight; the holistic reintegration of excessive investors in harm with presumption of equal moral worth; and degree to which our distributive evaluative ethic is co-determinative in the responsive process—all subject to genealogical evaluation of forces."

Our distributive justice principle (how rewards and burdens are distributed in society) is a composite of a number of positions in the literature. From Marx, we draw from his classic substantive principle "from each according to his abilities, to each according to his needs"; from May (1995), the notion of "antirepresentationalism";[31] from Deleuze, Guattari, Spinoza, and Nietzsche, "desire," joyful passions and promotion of differences; and from Nietzsche, the notion of evaluation of forces. A reasonable start in offering further guidance for "genealogical evaluation of forces" is provided by Fraser (1997, 2000; see also Woolford's comments, 2009) in her three-part criteria each of which is either more affirmative or more transformative. The first is *recognition*, a reversal of practices of misrecognition leading to exclusion, and it has two expressions: *affirmative recognition*, although recognizing harms of reduction, recognition does not go as far as recognizing institutional forms of harms; *transformative recognition* goes further to challenge cultural and symbolic forms. The second is *redistribution*, a reversing of economic injustices that lead to disenfranchisement, marginality, and it has two forms: *affirmative recognition* deals with superficial compensatory actions; *transformative redistribution* concerns addressing institutional-wide economic forms of diminishment and exploitation. Third is *representation*, which concerns voice provided for otherwise disenfranchised, and their two forms are: *affirmative representation* merely provides minimal voice representation, but short of system challenging; *transformative representation* deals with a socius-wide enhanced distribution of opportunities for voice to all members, in tune with May's (1994) call for antirepresentationalism. Fraser's model can also be reconceptualized along deterritorialization dynamics, which we will develop below. Woolford tells us Fraser's three-part model acting as "guideposts" recognizes more superficial or

31. Not representing others in their behalf and providing opportunities in which they can speak in their own behalf.

more holistic structural approaches and thereby better evaluates responses to injustice.

Another approach supportive of our two principles has been expressed by Gibson-Graham (2006). After reviewing types of "communities" (pp. 85–87) and suggesting many of these advocate a becoming-of-the same, they suggest the term "community economy" that stresses becoming. Their "ethical practice" (ibid., 88) would consist of "cultivating an awareness of": "[1] what is necessary to personal and social survival; [2] how social surplus is appropriated and distributed; [3] whether and how social surplus is to be produced and consumed; and [4] how a commons is produced and sustained." They are cognizant that in their model of a socius under continuous metamorphosis, contradictions and dislocations are ever present, but suggest "the individual needs nourishment and encouragement from without to sustain acts of self-cultivation, to see changing selves as contributing to changing worlds" (ibid., 162). This is a continuous process of disidentifying with dominant molar discourses and identifications and investing in becoming practices and emergent molecular forms (see also ibid., 77). It is in this spirit that we can consider Hardt and Negri's (2009: 282) call for a "common wealth, "free access, free use, free expression, free interaction—that stands against private control, that is the control exerted by private property, its legal structures, and its market forces."[32] And with the ascendancy of becoming-identifications in emergent community economies, Hardt and Negri (2009: 311) argue, a threshold would be reached at which point we mutate to becoming-others in an ubiquitous common wealth.

Deleuze and Guattari (1987: 52–56, 508–510) developed a four-part model[33] by which we could evaluate the deterritorialization component of assemblages.[34] We want to offer some preliminary integration with quantum holography. Recall, from Chapter 3, all entities can be conceptualized as assemblages that are either more molar or molecular in form. Each assemblage, in turn, is composed of two orthogonal dimensions: along one is content—expression, along

32. Hardt and Negri (2009: 309) thus advocate a maximum variability in flux, a continuous variation. "Freedom of movement," they argue, "would thus configure a freedom of space, allowing the multitude to flow to where it can be the most creative, organize the most joyful encounters, and establish the most productive relationships."

33. Deleuze and Guattari did not fully develop this four-part model, but have scattered passages throughout their text where the core elements are explained. In the final pages (1987: 508–510) they offer a few suggestions on this model. Patton (2000) has developed it more fully and we are inspired by his work.

34. Recall, Delanda (2006) has suggested that a person can be also conceived of as an assemblage; so, too, networks, groups, interactions, community, states, etc.

the other, deterritorialization—reterritorialization. The specific instantiations of their interactions produce distinct "lines of flight," a logic, a specific manifestation of the operative abstract machine of each form of assemblage. With this in mind, Deleuze and Guattari explain four ways of evaluating existing assemblages. Along one dimension, there lies the difference between relative and absolute deterritorialization. Relative forms describe restrictive movements[35] in the actual striated space (e.g., Euclidean geometric coordinate systems); whereas absolute forms describe changes in the virtual realm, the plane of consistency, unformed matter, and "smooth space," (e.g., Hilbert space of interacting dynamic vectors, and fractal geometric space indicating the infinity of in-between dimensions) (ibid., 56). Each form can further be seen in terms of positive or negative.

Negative forms of relative deterritorialization are where the reterritorialization process leads to capture, obstructions, co-optation, rigid codification, closure, stasis, stratification; positive is where it leads to the connections of lines of flight within the actual while leaving the virtual undisturbed, that is, core assumptions and overall institutional arrangements are unaffected. Negative forms of absolute deterritorialization lead to reconfigurations in the virtual, the plane of immanence, toward unformed matter that is readily subject to capture and axiomatization. They lead to "lines of destruction or death" (ibid., 510).

Positive forms of absolute deterritorialization lead to lines of flight that produce ongoing diverse connections of various lines of flight, "revolutionary connections" (ibid., 473). That is, a continuous decoding, becoming, creation, mutation, and metamorphosis are assured by connections that are not brought within an overall axiomatic system, but remain dynamic, emergent, and dissipative. They refer to this as a "new earth," a new socius. Historical practices could thus be measured and evaluated by what form of deterritorialization is at play. Consider, for example, the recent insurgent activities in North Africa, often termed the "Arab Spring." Examining each country could be along the lines of this evaluative schema. Egypt, for example, has undergone change that initially witnessed materializations more along the lines of positive absolute deterritorialization, but more recently has reverted to earlier forms of repression, or negative deterritorialization. A recent book by Olsen *et al.*, *Transitional Justice in Balance* (2010: 16), an admirable and productive thrust in

35. Deleuze and Guattari (1987: 509) define "movement" in terms of diverse bodies in relation to speed of movement. This concerns the speed of deterritorialization (ibid., 53), as they argue: "a function of differential speeds, relations, and rates" (ibid., 54).

empirically investigating global changes in repressive regimes and forms of justice instantiated, offers a too limiting standard of evaluations: "strengthening democracy" and "reducing human rights violations."[36] Deleuze and Guattari's model could, in a worthy exercise, be further integrated to measure "success."

We can see how Fraser's "affirmative recognition" is consistent with Deleuze and Guattari's (1987) concept of a positive relative deterritorialization, whereas "transformative" is more akin to a positive absolute deterritorialization. Deleuze and Guattari provide a theoretical framework by which affirmative or transformative practices could be understood. Quantum holography, however, can provide further specification and a physicality to information construction and processing. Returning to Bradley's model of communication (Figure 4), molecular assemblages and positive absolute deterritorialization could be situated in phase space in the lower right hand corner of low to medium control/high flux. Recall from Bradley (1998: 130), "the interpenetration between flux and control operates as a communicative-processing network that in-forms the transformation of potential energy into collective work." Looking at the "expression" dimension of assemblages, we can see how a resonating signature wave begins to emerge (i.e., Deleuze and Guattari's "refrain") reflective of the dynamic, open, and mutating field. Within this field, new images are being constructed (semiotics),[37] and a new language (linguistics)[38] consistent with this, always potentially subject to capture and axiomatization. Commonalities within the socius "establish a bandwidth of behavior and ... an energetic resonance or vibration—a vibe—that characterizes the group" (Bradley, 2011: 206). This is a "socio-bio-emotional bandwidth of energy frequency to which all members are attuned" (ibid.). Thus, this is a constancy, much like the "social wave function" pointed out by Wendt (2006) with which resonance can be established via phase conjugation. It is also the basis of a generalized community other. Every occasion of human interaction is the basis of constructions of logons (intersections of spectra and spacetime patterns), quanta of information that form the basis of perception-images, subsequent action-images, and motoric action. These interference patterns, or spectra, encode holographic forms of information that are nonlocally stored in the holofield/ZPF. We also see how

36. They also limit themselves to forms of responses to harms: trials, truth commissions, amnesties, reparation and lustration practices, although, ultimately they do encourage a "holistic approach" that draws from each.

37. This comports with Deleuze's (1986, 1989) cinema studies in developing a description of types of images and a taxonomy of signs that reflect them.

38. Linguistics defined by how the perpendicular axes of paradigm (lexicon, semantics) and syntagm (syntax, grammar, codes) define precise ways of constructing narrative.

a more molecular or molar materialization of the dynamics of schema QD resonates with this field. Lines of flight, or affordances, within this milieu are encoded frequency patterns with which a person engages. This milieu, we previously argued, is a vector field of forces, criss-crossing frequency patterns encoding holographic information. Bradley's version of a "space-time constrained hologram," or a logon, is the end result of the phase conjugation dynamic we outlined in our discussion of schema QD in Chapter 4.

A developing emergent "new earth," characterized by a more crystalline structure of the regime of signs, will be an emergent as it responds to the plethora of novel images appearing on the plane of immanence. We envision an emerging vector field reflecting new forms of images of mutating socio-economic relations of late modernity/postmodernity; a semantic space reflecting a regime of signs more crystalline in structure; a milieu in which the community generalized other is both a basis of "we" feelings, of solidarity, but also of recognition of differences and of becoming-other; and the instantiation of an agent in-process who draws from these resources in constructing social reality. Certainly greater differentiation is the basis of both appreciating the differences in others, but also of ambiguities, tensions and conflict to which a truly transformative justice needs to be created (see, for example, Unger's notion of "superliberalism" and new forms of tensions assumed, 2004). The mobile cuts of spacetime, or perception-images, will reflect interconnections, becoming-other, becoming-minortarian. It is also within this space we envision the actualization of "joyful passions," and the increase of power to effect and be affected. Our distributive and retributive principles of justice can be seen as "guideposts" in evaluating these developments.

Bradley (1998: 131: 2010: 151–153) has also offered how new community-shared holographically encoded information emerges. He argues, in his earlier model that instabilities between flux and control produce a new impetus toward constructions of quantized information and its transmission. Recall, his ideal group (attractor basin) is situated along a diagonal of Figure 4 located in phase space in the middle of his diagram, or the upper right hand corner where transformations and novelty prevails. The middle region in our view, however, represents more homeostatic dynamics, centripetal forces reconstituting the same. We have argued, rather, that a phase space located at the low control/high flux is consistent with molecular assemblages and a positive absolute deterritorialization. We view the identified "transformation and novelty" phase space much too inviting for the forces of capture, coding, and axiomatization, and, instead, locate it in the lower right hand area.

In his later works (2010: 151–153), Bradley introduces a "feedforward model" which operates much like a temperature dial that once reset, the new temperature is continuously maintained. In his model, everyday constructions of logons are matched against the distinct signature wave of the group, and where discordant (i.e., non-resonance), "pressure is created as feedback to the group to update or even reform its basic sociocultural organization" (ibid., 153). Once changed, a new "holographic signature" is created by which new logons are matched. Where matching exists, harmonic resonance, and information communication takes place. What is still needed here, however, is the work of the inter- and intra-subjectively constituted subject.

Mitchell and Staretz's (2011: 14) model of change is an advancement, even though it does need augmentation with a quantum holography-informed social organizational theory such as a version of Bradley's developed in this book. In their model, "inputs," or received quantum holographic waves—envisioned as affordances that crisscross a vector field—initiate phase conjugation dynamics, but a pause exists in immediate response by the percipient during which memory images from the ZPF are retrieved and evaluated. Here we interpose Bergson, Deleuze, and our schema QD in suggesting how recognition-images are summoned, brought within quantum entanglement, and at thresholds of intensity, collapse into one instantiation,[39] that becomes the virtual outgoing signature wave forming the basis of a circumscribed reconstruction of encounters in three-dimensional space, now seen as reality. It is to this returning constituted image that the percipient now consciously responds, even though, more unconscious backward time referral of quantum information has started the response earlier than the conscious recognition of a person's actions. In molecular assemblages and positive forms of absolute deterritorialization, these constructions are always in process, always subject to falsifying narratives, the "power of the false." In negative forms of deterritorialization, reactive forces, and sad passions, assure the ascendancy of the "man of truth" and axiomatization.

A transformative justice would build on and be mindful of the plentiful critiques of restorative justice. It is not to suggest that restorative justice is not well intended, but only to underscore the need to resist the forces of axiomatization and negative forms of deterritorialization/reterritorialization that are often a consequence or possibility of this approach. An affirmative nomadic transformative justice would be a holistic approach that builds on the work of Holland (1999, 2011), Hardt (1993), Unger (2004), and Hardt and Negri (2001, 2005, 2009) in providing context for the development of societal-wide molecular assemblages,

---

39. Recall from Chapter 4 that a complex interaction among the various dyads in schema QD shape perception-images and with its actualization, becomes a momentary, relatively stabilized matrix generating a nuanced signature wave.

and in providing further impetus for those advocating reformulation of restorative justice, such as theorized by Woolford (2009) and Pavlich (2005).

A particularly inspiring national proposal for a transformative justice was recently suggested by a working group of scholars in Jamaica (Chevannes, 2001; Milovanovic, 2012). One central idea was the establishment of a "Peace Institute." Their suggested composition would include a cross section of disciplines from various universities and colleges, professional organizations, and religious groups who would engage indigenous grassroots organizations, community leaders, and affected populations. Assuming an inherent link between crime and social structure, the proposal (see Chevannes, 2001) calls for a "transformative change" that would include three aims: [1] "facilitate dialogue at different levels and sectors of the society that are in conflict," [2] "help develop and enhance indigenous peacemaking capacities," and [3] "facilitate peace-oriented development work amongst grassroots organization." To this end they (ibid.) provide seven specific goals:

> "i. provide scholars, activists and other community leaders a common place to engage in research (including development of relevant methodologies), disseminate knowledge, and develop pedagogies around issues of community and communal peace
> ii. ground and connect itself with indigenous organizations around the country, in the region and internationally that are working "on the ground" settling disputes and developing programs of community-centered rehabilitation, peacemaking and restorative justice
> iii. observe, monitor and report, through mass information outlets, on national and regional achievement, or movement towards, goals and transformative justice
> iv. facilitate and/or offer its resources (physical space, staff, funding, communications technology) to groups enjoined in combat, yet desiring peace
> v. sponsor public workshops, forums, colloquia and seminars on and dramatizations of topics relating to peace, justice, responsibility and civil society
> vi. publish its own peacemaking journal, which would attract submissions from the region and the wider world
> vii. encourage and promote the work of its scholars."

This proposal for a transformative justice, with further adaptations for regional variations, could certainly be infused with some evaluative component previously suggested. Our offered distributive and retributive justice principles could be grounds for further discussion, refinement, and transformation.[40] Deleuze and Guattari's suggestions for evaluation of the forms of deterritorialization taking place could be guidelines. A quantum holography integration would suggest the necessity to reconsider simple understandings of causation, responsibility, reality construction, and information transmission. Ultimately, the measure would be: to what degree is the emerging socius and institutions (assemblages) supporting and encouraging the increase of the power to effect and be affected (Spinoza); to what degree are community economies and a common wealth being established; to what degree are singularities, which we are, respected and cultivated; and to what degree are harms of reduction and repression abating. All this subject to a genealogical analysis of forces[41] in historical moments.[42]

We would also suggest that given the mobilization of these elements that a crystalline regime of signs (Deleuze, 1988, 1999) would be engendered along with molecular assemblages, a quantum holographic lexicon that incorporates a wordspace of associative word linkages reflecting interconnections, and a "creative evolution" (Bergson, 1998). Ultimately, a transformative justice, as well argued by Woolford (2009: 148), and well in line with the work of Deleuze and Guattari, is a work in process, resisting static (molar) conceptualizations.[43]

---

40. For example, it could include a modified Parker's (1999) call for a plethera of "access to justice plans" whereby all organizations (public and private) ongoingly reflectively assess (genealogical method!) to what degree are improvements to access to justice taking place (i.e., diminishment of harms of reduction and repression).

41. It is here where the intellectual can offer a constructive component to social struggles. Consider Foucault (1975: 62): "the intellectual no longer has to play the role of an advisor. The project, tactics and goals to be adopted are a matter for those who do the fighting. What the intellectual can do is provide the instruments, and at the present time, this is the historian's essential role. What's effectively needed is a ramified, penetrative perception of the present ... a topological and geological survey of the battlefield—that is the intellectual's role."

42. My keynote address (Milovanovic, 2012) for the conference held in Jamaica, "Crime and Social Policy in the Caribbean," was attended by a broad representation of activists, scholars, practitioners, students, community members, media, and others, who after listening attentively to a rather highly abstract presentation, were to subsequently engage in highly provocative discussions. The organizing committee knew in advance that my presentation would be highly abstract and moving into quantum theory, but encouraged and welcomed it because following the same traditional course in crime fighting in Jamaica and surrounding Caribbean islands had dismal, failing results.

43. May (1994: 112, see also 112–119), drawing from Deleuze, Lyotard, and Foucault, suggests some key "general guidelines" which "include the call for social, personal, and political experimentation, the expansion of situated freedom, the release of subjected discourses and genres, and the limitation and reorientation of the role of the intellectual." There is much compatibility with Unger's (2004) notion of an "empowered democracy."

Conclusion

Our final chapter has applied quantum holography to transformative justice. As we have seen, nothing short of a holistic approach is necessary for transformative justice practices to grow and develop, necessitating a rethinking of the type of social formations which currently exist. These suggestions are more in the spirit of beginning a discussion in an alternative direction that challenges the classical-materialist paradigm that has strangled innovative thinking in criminology, law, and social justice. Quantum holography has provided an alternative way of looking at information construction, storage, transmission, and interpretation. This is not to dismiss much innovative process-informational thinking already being currently conducted, particularly inspired by Deleuze and Guattari. Nor is it to dismiss the argument that both classical as well as quantum principles are at work in the "real" world; nor, on the other hand, that the quantum realm is irrelevant to the everyday macro-level societal interactions. We concluded with a call for developing alternative forms of evaluation of the types of institutions (assemblages) within which we are situated. Undoubtedly, some will see this as too abstract for the real world of crime, drugs, gangs, and prisons. But we must rise to the occasion and re-think core ontological assumptions of current theorizing and practices to break out of the chains with which we have enveloped ourselves.

Chapter 7

Conclusion

We have just begun our journey. There is much to be done. We are entering a new era of quantum holography theory. In short time we look forward to quantum computers and holography-based discs that will revolutionize not only information storage and processing, but will provide a model which will bring in from the back door a re-thinking of Newtonian based ontologies. Quantum theory firmly established in the 1920s has led to much technological application. Holography theory having developments in the 1940s and 1990s has revised our perspectives on the nature of the cosmos. Scholarly studies have only recently begun in applying quantum holography in the social and neurobiological sciences. It hasn't taken off, due, of course, to the shackles with which we are enchained. We have ventured forth to provide some provisional suggestions on how the emerging process-informational paradigm may apply in criminology, law, and transformative justice. There is now the possibility of developing a bona fide notion of an inter- and intra-subjectively constituted agency which recognizes an alternative method in creating images and social reality. Phase conjugation was developed to indicate how information can be encoded, stored, transmitted, and interpreted.

There is much current thought on the nature of the holographic principle. We developed four of the more prominent versions. The most radical of the four even suggests that perhaps we are all holograms in a holographic-based cosmos, projections from an enclosed surface area on which all information is stored. Henri Bergson in 1896 was already suggesting a holographic cosmos in his *Matter and Memory*. As outlandish as it may sound, there is much serious research in cosmology and physics investigating this possibility. If this turns out to be supported, it is hard to fathom what this would do to our fundamental understandings of the cosmos and being human. Meanwhile, three other versions of the quantum holography principle are suggestive as to incorporation in the social sciences. Fortunately, we have had innovative thinkers who have dared to think differently lead the way. To name a few of the most prominent, David Bohm, Karl Pribram, Denis Gabor, Henry Stapp, Roger Penrose, and Stuart Hameroff. In applications to the social sciences, key figures are Raymond Bradley, Stephen Robbins, and Alexander Wendt. The work by Peter Marcer, Walter Schempp, Ervin Laszlo and Edgar Mitchell has provided the fundamentals of how information is encoded, stored, transmitted, and interpreted. There exists a recently started journal *NeuroQuantology* that is a leading outlet for studies on quantum holography theory. We also witness an increasing number of empirical studies in the area of quantum cognition focusing on the wherewithal of the human lexicon.

This book has been influenced substantially by the insights developed by Gilles Deleuze and Felix Guattari. They have provided us with a fresh vision on the potentials for increasing capacities of the person to effect and be affected. Their notion of becoming, a becoming-other, rather than being offers us a refreshing view on the nature of being human. We have seen that integration with quantum holography can be a productive direction in rethinking our basic ontological premises. One looks far and wide for publications in the dominant journals of criminology and law for articles that are based on their work. It is an unfortunate state of affairs. It allows abstract empiricism and positivism to flourish devoid of innovative and necessary theory. Law, too, with its narrow, linear logic of formal rationality is but a relic of questionable assumptions. There has been some inclusion of classical probability theory to bring these studies into a less linear, deterministic framework. But a quantum probability has emerged that better situates the emerging sciences. It has, however, yet to be taken seriously in criminology.

The field of criminology is in need of reformulations based on the process-informational paradigm. Law, too, has found itself constrained by its ontological assumptions rooted in the classical-materialist paradigm. Law school and legal practices have become bastions of limited thinking, and too often, supporters of the status quo. And the development of a more holistic transformative justice awaits based on the new paradigm. Much needs to be done. Ontologies need to be rethought. We need to dare to suspend our ontological premises and venture forth in developing and applying a process-informational paradigm.

References

Acorn, A.E. (2004) *Compulsory Community*. Minneapolis: University of Minnesota Press.

Adler, F. (1956) *The Individual Psychology of Alfred Adler*. Eds. H. Ansbacher and R. Ansbacher. New York: Harper and Row.

Agamban, G. (1993) *The Coming Community*. Minneapolis: University of Minnesota Press.

Aharonov, Y. and M.S. Zubairy (2005) "Time and the Quantum Erasing the Past and Impacting the Future." *Science* 307: 875–879.

Alley, C.O, O. Jakubowicz, W.C. Wickes (1987) "Results of the Delayed Random Choice Quantum Mechanics Experiment With Light Quanta and Proposal of a New Type of EPR Experiment Using Light Quanta Produced by a Nonlinear Optical Process," In Namiki (Ed.), *Proceedings of the 2nd International Symposium: Foundations of Quantum Mechanics in the Light of New Technology*. Tokyo, Physical Society of Japan, pp. 36–52.

Anderson, E. (1999) *Code of Ethics*. New York: W.W. Norton and Company.

Arrigo, B. (1998) "Reason and Desire in Legal Education." *International Journal for the Semiotics of Law* 11: 2–24.

Arrigo, B. and D. Milovanovic (2009) *Revolution in Penology: Rethinking the Society of Captives*. New York: Rowman and Littlefield Publishers, Inc.

Aspect, A., Dalibard, and G. Roger (1982) "Experimental Test of Bell's Inequalities Using Time-Analyzing." *Physical Review Letters* 49(25): 1804–1807.

Austin, J.L. (1975) *How To do Things with Words*. Harvard University Press.

Bajpai, R.P. (1998) "Coherent Nature of Biophotons: Experimental Evidence and Phenomenological Model." In J. Chang, J. Fisch, and F.A. Popp (Eds.) *Biophotons*, pp. 323–339. Kluwer: Academic Publishers.

Barbour, J. (2001) *The End of Time: The Next Revolution in Physics*. New York: Oxford University Press.

Barr, S. (1964) *Experiments in Topology*. New York: Dover Publications.

Basharpoor, S., M. Narimani, H. Gamari-Give, A. Abolgasemi, and P. Molavi (2011) "Effect of Cognitive Processing Therapy and Holographic Reprocessing on Reduction of Posttraumatic Cognition in Students Exposed to Trauma." *Iranian Journal of Psychiatry* 6(4): 138–144.

Batiz, Z. and B. Chauhan (2009) "Holographic Principle and Quantum Physics." *NeuroQuantology* 7(4): 665-676.

Baudrillard, J. (1981) *For a Critique of the Political Economy of the Sign*. St. Louis: Telos.

Baudrillard, J. (1983) *Simulations*. New York: Semiotext(e).

Bauman, Z. (2000) *Liquid Modernity*. Cambridge: Polity Press.

Beck, U. (1992) *Risk Society*. London: Sage.

Beck, U. (2008) *World at Risk*. Cambridge: Polity Press.

Bekenstein, J. (2003) "Information in the Holographic Universe." *Scientific American* 289(2): 58–66.

Bem, D.J. (2011) "Feeling the Future: Experimental Evidence for Anomalous Retroactive Influences on Cognition and Affect." *Journal of Personal Social Psychology* 100(3): 407–425.

Bem, D.J., J. Utts, and J. Johnson (2011b) "Must Psychologists Change the Way They Analyze Their Data?" *Journal of Personal Social Psychology* 101(4): 716–719.

Benhabib, S. (1986) "The Generalized and the Concrete Other." *Praxis International* 5: 402–24.

Benjamin, W. (1986). "Critique of Violence." In Schocken (Ed.). *Reflections: Essays, Aphorism, Autobiographical Writings*, pp. 277–300. Schocken.

Benioff, P. (2003) "Language is Physical." *Quantum Information Processing* 1(6): 495–509.

Benkler, Y. (2006) *The Wealth of Networks: How Social Production Transforms Markets and Freedom*. New Haven, CT: Yale University Press.

Benveniste, E (1971) *Problems in General Linguistics*. Coral Gables: University of Miami Press.

Berger, P. and T. Luckman (1966) *The Social Construction of Reality*. Anchor Books.

Bergson, H. (1998) *Creative Evolution*. London: MacMillan Co.

Bergson, H. (1999) *Duration and Simultaneity*. Clinamen Press, Ltd.

Bergson, H. (2002) *Matter and Memory*. New York: Zone Books.

Bergson, H. (2007) *Mind-Energy*. London: University of London Press.

Bersot, H. and B. Arrigo (2010) "Inmate Health Issues, Solitary Confinement, and Cruel and Unusual Punishment: An Ethical Justice Policy Inquiry." *Journal of Theoretical and Philosophical Criminology* 2(3): 1–82.

Beyleveld, D. (2001) *Human Dignity, in Bioethics and Law*. New York: Oxford University Press.

Blood, C. (2009) "Difficulties with Collapse Interpretations of Quantum Mechanics." *NeuroQuantology* 7(2): 302–313.

Bogue, R. (2003a) *Deleuze On Cinema*. New York: Routledge.

Bogue, R. (2003b). *Deleuze on Literature.* New York: Routledge.

Bohm, D. (1983) *Wholeness and the Implicate Order*. New York: Routledge.

Bohm, D. and D. Peat (1987) *Science, Order and Creativity*. New York: Bantam Books.

Borsellino, A. and T. Poggio (1972) "Holographic Aspects of Temporal Memory and Optomotor Responses." *Kybernetik* 10(1): 58–60.

Bourdieu, P. (1992). *The Logic of Practice*. Redwood City: Stanford University Press.

Bousso, R. (1999) "Holography in General Space-Times." arXiv:hep-th/99006022v2.

Bousso, R. (2002). "The Holographic Principle." *Review of Modern Physics* 74: 825–874.

Bowser, J. (2012) *Gangster Rap and Its Social Costs: Exploiting Hip Hop Using Racial Stereotypes to Entertain Americans*. Amherst, New York: Cambia Press.

Bracewell, R. (1989) "The Fourier Transform." *Scientific American*, (June): 62–69.

Bracher, M. (1988) "Lacan's Theory of the Four Discourses." *Prose Studies* 1: 32–49.

Bracher, M. (1993) *Lacan, Discourse and Social Change: A Psychoanalytic Cultural Criticism*. New York: Cornell University Press.

Bradley, R. (1998) "Quantum Vacuum Interaction and Psychosocial Organization," in D. Loyle (ed.) *The Evolutionary Outrider*. Pp. 117–149. Westport, Connecticut: Praeger.

Bradley, R. (2006) "The Psychophysiology of Entrepreneurial Intuition," Proceedings of the 3rd AGSE International Entrepreneurship Research Exchange. Auckland, New Zealand.

Bradley, R. (2007) "A Psychophysiology of Intuition: A Quantum Holographic Theory of Intuition." *World Futures* 63(2): 61–97.

Bradley, R. (2009). "Transformational Dynamics of Entrepreneurial Systems: Organizational Basis of Intuitive Action." *International Journal of Entrepreneurship and Small Business* 11(2): 183–204.

Bradley, R. (2010) "Detecting the Identity Signature of Secret Social Groups: Holographic Processes and the Communication of Member Affiliation." *World Futures* 66: 124–162.

Bradley, R. (2011) "Communication of Collective Identity in Secret Social Groups." *Behavioral Sciences of Terrorism and Political Aggression* 3(3): 198–224.

Braithwaite, J. (2002) *Restorative Justice and Responsive Regulation.* Oxford: Oxford University Press.

Braund, M. (2008) "The Structure of Perception: An Ecological Approach." *Kritike* 2(1): 123–144.

Bruza, P., K. Kitto, D. Nelson, and K. McEvoy (2008) "Entangling Words and Meaning," Association for the Advancement of Artificial Intelligence (www.aaai.org).

Bruza, P., K. Kitto, D. Nelson, and C. McEvoy (2009a) "Is There Something Quantum-Like About the Human Lexicon?" *Journal of Mathematical Psychology* 53(5): 363–377.

Bruza, P., K. Kitto, D. Nelson, and C. McEvoy (2009b) "Extracting Spooky-Activation-at-a-Distance from Considerations of Entanglement." arXiv:0901.4375v1.

Bruza, P. and R. Cole (2006) "Quantum Logic of Sematic Space: An Exploratory Investigation of Context Effects in Practical Reasoning." arXiv:quant-ph0612178v1.

Bruza, P., J. Busemeyer and L. Gabora (2009) "Introduction to the Special Issue on Quantum Cognition." *Journal of Mathematical Psychology* 53: 303–305.

Buber, M. (1971) *I and Thou.* New York Charles Scribner's Sons.

Bunge, M. (1959). *Causality and Modern Science.* N.Y.: Dover Publications.

Busemeyer, J. and P. Bruza (2012) *Quantum Models of Cognition and Decision.* Cambridge University Press.

Butler, J. (1993) *Bodies That Matter: On the Discursive Limits of Sex.* New York: Routledge.

Capek, M. (1961) *The Philosophical Impact of Contemporary Physics.* New York: D. Van Nostrand.

Capek, M. (1971). *Bergson and Modern Physics.* New York: Humanities Press.

Carter, J.S. (1993) *How Surfaces Intersect in Space: An Introduction to Topology.* New Jersey: World Scientific.

Castells, M. (2012) *Networks of Outrage and Hope: Social Movements in the Internet Age.* Cambridge, UK; Polity Press.

Chalmers, D. (1996) "Facing Up to the Problem of Consciousness." *Consciousness Studies* 2(3): 200–219.

Chalmers, D. (1997) *The Conscious Mind.* Oxford University Press.

Chang, J., J. Fisch, and F.A. Popp (Eds.) (1998). *Biophotons.* Dordrecht: Kluwer Academic Publishers.

Chardin, P. (2002) *The Heart of Matter.* Mariner Books. "Cease Fire Chicago." Renamed "Cure Violence." http://cureviolence.org/.

Chevannes, B. (2001) "Crime, Peace, and Justice For Jamaica: Jamaicans for Justice." Working paper can be found at: jamaicansforjustice.org.

Chuchland, P. (1986) *Neurophilosophy*. Cambridge, MA: MIT Press.

Cloward, R. and L. Ohlin (1966). *Delinquency and Opportunity*. New York: The Free Press.

Cohen, A. (1955) *Delinquent Boys*. New York: The Free Press.

Cornell, D. (1998) *At The Heart of Freedom*. Princeton: Princeton University Press.

Cottee, S. and K. Hayward (2011) "Terrorist (E)motives: The Existential Attractions of Terrorism." *Studies in Conflict and Terrorism* 34: 963–986.

Cox, G.E., G. Kachergis, G. Recchia, and M. Jones (2011) "Toward a Scalable Holographic Word-Form Representation." *Behavior Research Methods* 43(3): 602–15.

Cramer, J. (1986a) "The Transactional Interpretation of Quantum Mechanics." *Review of Modern Physics* 58: 647–688. Available online at: http://mist.npl.washington.edu/npl/int_rep/tiqm/TI_toc.html.

Cramer, J. (2001) "The Plane of the Present and the New Transactional Paradigm of Time." http://arxiv.org/ftp/quant-ph/papers/0507/0507089.pdf.

Cramer, J. (2006a) "Reverse Causation and the Transactional Interpretation of Quantum Mechanics." In D. Sheehan (ed.) *Frontiers of Time: Retrocausation–Experiment and Theory*, pp. 20–26. Melville, N.Y.: AIP Conference Proceedings, V. 863.

Cramer, J. (2006b). "The Blind Men and The Quantum." http://www.niac.usra.edu/files/library/meetings/fellows/mar06/Cramer_John.pdf.

Crick, F. (1994) *The Astonishing Hypothesis: The Scientific Search for the Soul*. New York: Touchstone.

Crick, F. and C. Koch (1995) "Why Neuroscience May be Able to Explain Consciousness." *Scientific American* 273: 84–85.

Crick, F. and C. Koch (1998) "Consciousness and Neuroscience." *Cerebral Cortex* 8(2); 97–107.

Crick, F. and C. Koch (2002) "A Framework for Consciousness." http://codatest4.library.caltech.edu/26/1/438.pdf.

Critchley, S. (1999) *The Ethics of Deconstruction: Derrida and Levinas*. West Lafayette, Indiana: Purdue University Press.

De Broglie, L. (1924) "A Tentative Theory of Light Quanta." *Philosophical Magazine* 35: 446–458.

De Vine, L. and P. Bruza (2010) "Semantic Oscillations: Encoding Context and Structure in Complex Valued Holographic Vectors." Quantum Informatics for Cognitive, Social and Semantic Processes: Papers from the AAA1, Fall Symposium (FS-10-88).

Delanda, M. (2002) *Intensive Science and Virtual Philosophy*. London: Continuum.

Delanda, M. (2006) *A New Philosophy of Society*. London: Continuum.

Deleuze, G. (1983) *Nietzsche and Philosophy*. New York: Athlone Press.

Deleuze, G. (1988) *Foucault*. Minneapolis, MN: University of Minnesota Press.

Deleuze, G. (1992) *Expressionism in Philosophy*. New York: Zone Books.

Deleuze, G. (2004) *L'Abécédaire de Gilles Deleuze, avec Claire Parnet*. Paris: DVD Editions Montparnasse.

Deleuze, G. (1986) *Cinema 1: The Movement-Image*. Minneapolis, MN: University of Minneapolis Press.

Deleuze, G. (1989) *Cinema 2: The Time-Image*. Minneapolis, M: University of Minneapolis Press.

Deleuze, G. and F. Guattari (1983) *Anti-Oedipus*. Minneapolis: University of Minneapolis Press.

Deleuze, G. and F. Guattari (1986) *Kafka: Toward a Minor Literature*. Minneapolis, MN: University of Minneapolis Press.

Deleuze, B. and Guattari (1987) *A Thousand Plateaus*. Minneapolis: University of Minnesota Press.

Deleuze and Guattari (1996) *What is Philosophy?* New York: Columbia University Press.

Dennett, D. (1991) *Consciousness Explained*. Boston, MA: Little Brown.

Denzin, N. (2007) *On Understanding Emotions*. Transaction Books.

Di Biase, F. (2009) "Quantum-Holographic Informational Consciousness." *NeuroQuantology* 7(4): 657–664.

DiCrisina, B. (1995) *Method in Criminology*. N.Y: Harrow and Heston.

Diric, P. (1930) *Principles of Quantum Mechanics*. New York: Oxford University Press.

Dor, J. (1992) *Introduction á la Lecture de Lacan*. Paris: Denoel.

Dyer-Witherford, N. (1999) *Cyber-Marx: Cycles and Circuits of Struggle in High-Technology Capitalism*. Chicago, Il: University of Illinois Press.

Eccles, J.C. (1985) "Mental Summation: The Timing of Voluntary Intentions by Cortical Activity." *Behavioral and Brain Sciences*, 8: 542–543.

Eddington, A.S. and J. Jeans (1984) *Quantum Questions: Mystical Writings of the World's Greatest Physicists*. New York: Random House.

Edelman, G. (1987) *Neural Darwinism*. New York: Basic Books.

Einstadter, W. and S. Henry (1995) *Criminological Theory: An Analysis of Its Underlying Assumptions*. Fort Worth, TX: Harcourt Brace College Publishers.

Einstein, A., B. Podolsky, and N. Rosen (1935) "Can Quantum-Mechanical Description of Physical Reality be Considered Complete?" *Physical Review* 47(10): 777–780.

Einstein, A. and O. Stern (1913) "Einige Argumente für die Annahme einer molekularen Agitation beim absoluten Nullpunkt". *Annalen der Physik* 40 (3): 551.

Einstein, A. (1995) *Ideas and Opinions*. New York: Broadway.

Elitsur, A. and S. Dolev (2006) "Multiple Interaction-Free Measurement as a Challenge to the Transactional Interpretation of Quantum Mechanics." In D. Sheehan (ed.) *Frontiers of Time: Retrocausation–Experiment and Theory*, pp. 27-43. Melville, N.Y.: AIP Conference Proceedings, V. 863.

Empey, L. and M. Stafford (1991) *American Delinquency: Its Meaning and Construction*. Belmont, CA: Wadsworth.

Epstein, S. (2014 *Cognitive-Experiential Theory: An Integrative Theory of Personality*. Oxford: Oxford University Press.

Erikson, K. (2004) *Wayward Puritans*. New York: Prentice Hall.

Ferrell, J. and C. Sanders (Eds.) (1995) *Cultural Criminology*. Boston: Northeastern University Press.

Ferrell, J. K. Hayward, and J. Young (1998) *Cultural Criminology*. London: Sage.

Fischer, C. (1982) *To Dwell Among Friends: Personal Networks in Town and City*. Chicago: University of Chicago Press.

Follet, M.P. (1949) *Freedom and Co-ordination*. London: Management Publications Trust.

Foucault, M. (1971) "Nietzsche, Genealogy, History," in D. Bouchard (Ed.), *Language, Counter-Memory Practice*. Ithaca: Cornell University Press.

Foucault, M. (1975) "Body/Power," in C. Gordon (Ed.) *Power/Knowledge*, pp. 55–62. New York: Pantheon Books.

Foucault, M. (1977) *Discipline and Punish*. London: Penguin.

Foucault, M. (1986) "Of Other Spaces." *Diacritics* 16: 22–27.

Foucault, M. (1994). *Ethics: Subjectivity and Truth*. P. Rabinow (Ed.). New York: The New Press.

Foucault, M. (2000) *Michel Foucault: Power*. J. Faubion (Ed.). New York: The New Press.

Foucault, M. (2003) *Society Must Be Defended*. New York: Picador.

Fraser, N. (1986) "Toward a Discourse Ethic of Solidarity." *Praxis International* 5: 427–429.

Fraser, N. (1997) *Justice Interruptus: Critical Reflections on the 'Postsocialist Condition*. New York: Routledge.

Fraser, N. (2000). "Rethinking Recognition." *New Left Review* 3: 107–120.

Freire, P. (2005) *Education for Critical Consciousness*. Continuum.

Freud, S. (2000) *The Interpretation of Dreams*. Wadsworth.

Gabor, D. (1946) "Theory of Communication." *Journal of the Institute of Electrical Engineers* 93: 429–441.

Gabor, D. (1948) "A New Microscopic Principle." *Nature* 161: 777–778.

Gabor, D. (1968) "Holographic Model of Temporal Recall." *Nature* 217: 1288–1289.

Gabor, D. (1971) "Holography, 1948–1971." Nobel Lecture, December 11, 1971.

Gabora, L. (2010). "The Emergence and Evolution of Integrated Worldviews." zrXiv:1001.1399v1.

Gailey, J (2007) "The Pro-Ana Culture as Edgework." Paper Presented at the American Society of Criminology Annual Meeting, November, Atlanta, GA.

Gallagher, S. (2008) "Direct Perception in the Intersubjective Context." *Consciousness and Cognition* 17: 535–543.

Gardenfors, P. (2004) *Conceptual Spaces: The Geometry of Thought.* MIT Press.

Garfield, J.L. (1995) *The Fundamental Wisdom of the Middle Way.* Oxford: Oxford University Press.

Garfinkel, A. (1981) *Forms of Explanation.* New Haven: Yale University Press.

Garfinkel, A. (1991) *Studies in Ethnomethodology.* London: Polity.

Gariaev, P.B. Birsthtein, A. Larochenko, P. Marcer, G. Tertishny, K. Leonova, and W. Kaempf (2001) "The DNA-Wave Biocomputer." *International Journal of Computing Anticipatory Systems* 10: 290–310.

Gariaev, P, M. Friedman and E. Leonova-Gariaev (2011) "Principles of Linguistic-Wave Genetics." *DNA Decipher Journal* 1(1): 1–24.

Gazzinga, M. (1985) *The Social Brain.* New York: Basic Books.

Gazzinga, M. (2012) *Who's in Charge? Free Will and the Science of the Brain.* New York, NY: Ecco.

Germine, M. (2008) "The Holographic Principle of Mind and the Evolution of Consciousness." *World Futures* 64: 151–178.

Gibson, J. (1966) *The Senses Considered as Visual Systems.* Boston: Houghton-Mifflin.

Gibson, J. (1979) *The Ecological Approach to Perception.* Hillsdale, N.J.: Lawrence Erlbaum.

Gibson, J. (1982) *Reasons for Realism.* Hillsdale, N.J.: Lawrence Erlbaum.

Gibson-Graham, J.K. (2006) *Postcapitalist Politics.* Minneapolis, MN: University of Minnesota Press.

Glassman, R. (1999) "Hypothesized Neural Dynamics of Working Memory." *Brain Research Bulletin* 50(2): 77–93.

Globus, G. (2007) "Mind, Matter, and Monad." *Mind and Matter* 5(2): 201–214.

Globus, G. (2010) "Toward a Quantum Psychiatry: Hallucination, Thought Insertion and DSM." *NeuroQuantology* 8(1): 1–12.

Goldstein, B. (1981) "The Ecology of J.J. Gibson's Perception." *Leonardo* 14(3): 191–195.

Graham, P. (Ed.) (2003) *Mary Parker Follet: Prophet of Management*. Washington, D.C.: Beard.

Granon-Lafont, J. (1985) *La Topologier Ordinaire de Jacques Lacan*. Paris: Point Hors Lignes.

Grass, F., H. Klima and S. Kasper (2004) "Biophotons, Microtubules and CNS: Is Our Brain a 'Holographic Computer?'" *Medical Hypotheses* 62: 169–172.

Greene, B. (2011) *The Hidden Reality*. New York: Alfred A. Knopf.

Greimas, A. (1987) *On Meaning*. Minneapolis, MN: Minnesota University Press.

Greimas, A. and E. Landowski (1990) "The Semiotic Analysis of Legal Discourse," in A. Greimas (Ed.) *The Social Sciences: A Semiotic View*, pp. 102–138. MN: Minnesota Press.

Grosz, E. (1994) *Volatile Bodies*. Bloomington, IN; Indiana University Press.

Gubrium, J. and J. Holstein (1997) *The New Language of Qualitative Method*. Oxford: Oxford University Press.

Halleck, S. (1967) *Psychiatry and the Dilemmas of Crime*. Berkeley: University of California Press.

Hameroff, S. (n.d.) "Consciousness, Whitehead and Quantum Computation in the Brain: Panprotopsychism Meets the Physics of Fundamental Spacetime Geometry." http://www.quantumconsciousness.org/whitehead.htm.

Hameroff, S. (1998) "Quantum Computation in Brain Microtubules? The Penrose-Hameroff 'Orch Or' Model of Consciousness." *Philosophical Transactions of Royal Society* 356: 1869–1896.

Hameroff, S. and R. Penrose (1996a) "Orchestrated Reduction of Quantum Coherence in Brain." *Journal of Consciousness Studies* 3: 36–53.

Hameroff, S. and R. Penrose (1996b) "Conscious Events as Orchestrated Space-Time Selections." *Journal of Consciousness Studies* 3(1): 36–53.

Hameroff, S. (2006) "Consciousness, Neurobiology and Quantum Mechanics." In J. Tuszynski (Ed.) *The Emerging Physics of Consciousness*, pp. 193–253. Berlin: Springer.

Hameroff, S. (2007a) "The Brain is Both Neurocomputer and Quantum Computer." *Cognitive Science* 31: 1035–1045.

Hameroff, S. (2007b) "Orchestrated Reduction of Quantum Coherence in Brain Microtubules." *NeuroQuantology* 5: 1–8.

Hameroff, S. (2010) "The 'Conscious Pilot'—Dendrite Synchrony Moves Through the Brain to Mediate Consciousness." *Journal of Biological Physics* 36: 71–93.

Hameroff, S. (2012) "How Quantum Brain Biology Can Rescue Conscious Free Will." *Frontiers in Integrative Neurosciences* 6: 1–17.

Hannagan, T., E. Dupoux, and A. Christophe (2010) "Holographic String Encoding." *Cognitive Science* 35: 79–118.

Hannagan, T., F. Dandurand, and J. Grainger (2011) "Broken Symmetries in a Location Invariant Word Recognition Network." http://www.neural-diaries.com/ThomasHannagan/Material/Hannagan,%20Dandurand,%20Grainger%20-%20NeCo%202011.pdf.

Hardt, M. (1993) *Gilles Deleuze*. Minneapolis: University of Minnesota Press.

Hardt, M. and A. Negri (2001) *Empire*. Cambridge, MA: Harvard University Press.

Hardt, M. and A. Negri (2005) *Multitude*. New York: Penguin Press.

Hardt, M. and A. Negri (2009) *Commonwealth*. Cambridge, MA: Harvard University Press.

Harris, H.D. (2002) "Holographic Reduced Representations for Oscillator Recall." In W.D. Gary and C.D. Schunn (Eds.) Proceedings of the 24th Annual Meeting of the Cognitive Science Society, pp. 423–428. Mahwah, NJ. Erlbaum.

Hawkins, D. (1985) "Black Homicide: The Adequacy of Existing Research for Devising Prevention Strategies." *Crime and Delinquency* 31: 83–103.

Hayes, R.P. (1994) "The Philosophy of Nagarjuna, Nagarjuna's Appeal." *Journal of Indian Philosophy* 22: 299–378.

Henry, S. and D. Milovanovic (1996) *Constitutive Criminology*. London: Sage.

Henry, S. and D. Milovanovic (1999) *Constitutive Criminology at Work*. New York: State University of New York Press.

Hochschild, J. (1983) *The Managed Heart: Commercialization of Human Feelings*. Berkeley, CA: University of California Press.

Hoffman, H. (2004) "Virtual-Reality Therapy." http://www.scientificamerican.com/article/virtual-reality-therapy/.

Hoffman, H. (2004) "Virtual-Reality Therapy." *Scientific American*.

Holland, E. (1999) *Deleuze and Guattari's Anti-Oedipus*. New York: Routledge.

Holland, E. (2011) *Nomad Citizenship*. Minneapolis: University of Minneapolis Press.

Huxley, T. (1986) *Method and Results: Essays*. Appleton.

Jackson, B. (1985) *Semiotics and Legal Theory*. New York: Routledge and Kegan Paul.

Jackson, B. (1995) *Making Sense in Law*. Liverpool, UK: Deborah Charles Publications.

Jackson, B. (1998) "Truth or Proof?": The Criminal Verdict." *International Journal for the Semiotics of Law* 33: 227–273.

Jacobs, J. (1961) *The Death and Life of Great American Cities*. New York: Radom House.

Jacques, V. E. Wu, F. Grosshans, F. Treussart, P. Grangier, A. Aspect, and J-F Rochl (2007). "Experimental Realization of Wheeler's Delayed-Choice Gedanken Experiment." *Science* 315: 966–968.

Jammer, M. (1954) *Concepts of Space*. Cambridge, MA: Harvard University Press.

Jenkins, H. (2008) "Gibson's 'Affordances': Evolution of a Pivotal Concept." *Journal of Scientific Psychology* (December): 34–45.

Jones, M. and D. Mewhort (2007) "Representing Word Meaning and Order Information in a Composite Holographic Lexicon." *Psychological Review*: 114(1): 1–37.

Jung, C. G. (2010) *Synchronicity*. Princeton University Press.

Kadar, E. and R. Shaw (2000) "Toward an Ecological Field Theory of Perceptual Control of Locomotion." *Ecological Psychology* 12(2): 141–180.

Katz, J. (1988) *Seductions of Crime*. New York: Basic Books.

Katz, J. (1999) *How Emotions Work*. Chicago: University of Chicago Press.

Katz, L. (2005) *Holographic Reprocessing*. New York: Routledge.

Katz, L., M.R. Snetter, A.H. Robinson, P. Hewitt, and G. Cojucar (2008) "Holographic Reprocessing: Empirical Evidence to Reduce Posttraumatic Cognitions in Women Veterans with PTSD from Sexual Trauma and Abuse." *Psychotherapy* 45(2): 186–198.

Kelly, M., D. Blostein and D. J.K. Mewhort (2013) "Encoding Structure in Holographic Reduced Representations." *Canadian Journal of Experimental Psychology* 67(2): 79–93.

Kelman, M. (1981) "Interpretive Construction in the Substantive Criminal Law." *Stanford Law Review* 33: 591–674.

Kelsey, A. (2013) "The Law of Physics and the Physics of Law." *Regent University Law Review* 25: 89–102.

Kennedy, D. (1997) *A Critique of Adjudication*. Cambridge: Harvard University Press.

Keyes, C. (2002) *Rap Music and Street Consciousness*. Chicago: University of Illinois Press.

Kim, Y.H., R. Yu, S.P. Kulik, Y.H. Shih and M. Scully (2000) "A Delayed Choice Quantum Erasure." *Physical Review Letters*: 84: 1–5. Also available: http://cds.cern.ch/record/381875/files/9903047.pdf.

Kitto, K. and P. Bruza (2011) "Meaning In Context," http://www.per.marine.csiro.au/staff/Fabio.Boschetti/Context/Context_KittoBruza.pdf.

Kitto, K., B. Ramm, L. Sitbon, and P. Bruza (2010) "Quantum Theory Beyond the Microscopic: Information in Context." http://eprints.qut.edu.au/38001/1/c38001.pdf.

Kitto, K., P. Bruza, and L. Gabora (2013) "A Quantum Information Retrieval Approach to Memory." srXiv:1310.300va.

Koch, C. (2004) *The Quest for Consciousness.* Englewood, CO: Roberts and Company.

Koch, C. and K. Hepp (2006) "Quantum Mechanics in the Brain." *Nature* 440: 611–612.

Koch, C. and K. Hepp (2007) "The Relation Between Quantum Mechanics and Higher Brain Functions: Lessons From Quantum Computation and Neurobiology." http://www.theswartzfoundation.org/papers/caltech/koch-hepp-07-final.pdf.

Kolak, D. (2004) *I Am You: The Metaphysical Foundations for Global Ethics.* Dordrecht: Springer.

Krohn, M. (1986) "The Web of Conformity: A Network Approach to the Explanations of Delinquent Behavior." *Social Problems* 33: 581–593.

Kubrin, C. (2005) "Gangsters, Thugs, and Hustlers: Identity and the Code of the Street in Rap Music." *Social Problems* 52(3): 360–378.

Kuttner, R. (2011) "The Negotiation Within: The Wave/Particle Tension in Negotiation." *Harvard Negotiation Law Review* 16: 331–355.

Kwiat, K.W., A.M. Steinberg, and R.Y. Chiao (1992) "Observation of a "Quantum Eraser": A Revival of Coherence in a Two-Photon Interference Experiment." *Physical Review* 45: 7729–7739.

Lacan, J. *Écrits* (1977). New York: Norton.

Lacan, J. (1991) *l'Envers de la Psychanalysis.* Paris: Editions du Seuil.

Lanier, M. and S. Henry (2010) *Essential Criminology.* Boulder, CO. Westview Press.

Laplace, P., (1951) *Philosophical Essay on Probabilities,* New York: Dover.

Laszlo, E. (1995) *The Interconnected Universe.* London: World Scientific Publishing Co.

Laszlo, E. (1991) *The Connective Hypothesis: Foundations of an Integral Science of Quantum, Cosmos, Life and Consciousness.* Albany, New York: State University of New York Press.

Laszlo, E. (2007) *Science and the Akashic Field.* Rochester, Vermont: Inner Traditions.

Lefebvre, A. (2008) *The Image of Law.* Stanford, California: Stanford University Press.

Levinas, E. (1969) *Totality and Infinity.* Pittsburgh, Pennsylvania: Duquesne University Press.

Levinas, E. (1998) *Entre Nous: Thinking of the Other.* New York: Columbia University Press.

Lewin, K. (2007) *Principles of Psychology.* Munshi Press.

Lewin, K. (2010) *Resolving Social Conflict and Field Theory in Social Sciences.* American Psychological Association.

Li, J. (2013) "A Timeless and Spaceless Quantum Theory of Consciousness." *NeuroQuantology* 11(3): 431–442.

Libet, B., E. Wright, B. Feinstein, and D. Pearl (1979) "Subjective Referral of the Timing for a Conscious Sensory Experience." *Brain* 102: 193–224.

Libet, B.E., C. Gleason, E. Wright and D. Pearl (1983) "Time of Conscious Intention to Act in Relation to Onset of Cerebral Activities (Readiness-Potential): The Unconscious Initiation of a Freely Voluntary Act." *Brain* 106: 623–642.

Libet, B.E. (1981) "The Experimental Evidence for Subjective Referral of a Sensory Experience Backward in Time: Reply to P.S. Churchland." http://web.gc.cuny.edu/cogsci/private/Libet-referral-backward-PS.pdf.

Libet, B.E. (1985) "Unconscious Cerebral Initiative and the Role of Conscious Will in Voluntary Action." *Behavioral and Brain Sciences* 8(4): 529–566.

Libet, B.E. (1992) "The Neural Time-Factor in Perception, Volition, and Free Will." *Review de Metaphysique et de Morale* 97: 255–272.

Libet, B.E. (2002) "The Timing of Mental Events: Libet's Experimental Findings and Their Implications." *Consciousness and Cognition* 11: 291–299.

Libet, B. E. (2004) *Mind Time: The Temporal Factor in Consciousness.* Harvard University Press.

Libet, B.E. (2006) "Reflections on the Interaction of the Mind and Brain." *Progress in Neurobiology* 78(3): 322–326.

Lippens, R. (1998) "Alternatives to What Kind of Suffering: Toward a Border-Crossing Criminology." *Theoretical Criminology* 2(3): 311–343.

Litowitz, R. (1997) *Postmodern Philosophy of Law.* Lawrence: University Press of Kansas.

Lois, J. (2005) "Gender and Emotion Management in the Stages of Edgework." In S. Lyng, *Edgework: The Sociology of Risk-Taking*, pp. 117–152. New York: Routledge.

Longuet-Higgins, H.C. (1968) "Holographic Model of Temporal Recall." *Nature* 217: 104.

Lopez, H. (1992). *Rebellious Lawyering.* San Francisco: Westview Press.

Lutz, A., Greischar, L., Rawlings, N., Ricard, M., and R. Davidson (2004) "Long-Term Meditators Self-Induce High-Amplitude Gamma Synchrony During Mental Practice." *Proceedings of the National Academy of Sciences*, USA, 1-1, pp. 16369–16373.

Lyng, S. (1990) "Edgework: A Social Psychological Analysis of Voluntary Risk Taking." *American Journal of Sociology* 95(4): 851–886.

Lyng, S. (2005) *Edgework: The Sociology of Risk-Taking.* New York: Routledge.

Lyotard, J. F., (1988) *Just Gaming*. Minneapolis: University of Minnesota Press.

Maldacena, J. (1998) "The Large N Limit of Superconformal Field Theories and Supergravity." *Advances in Theoretical mathematical Physics* 2: 231–252.

Maldacena, J. (2005) "The Illusion of Gravity." *Scientific American* 17: 57–63.

Manning, P. (1998). "Media Loops." In F. Bailey and D. Hale (Eds.) *Popular Culture, Crime and Justice*. Belmont, CA: Wadsworth.

Marcer, P. (1995) "The Need to Define Consciousness—A Quantum Mechanical Model," Symposium, pp. 23–45. P.Marcer and A. Fedorec (Eds.) University of Greenwich.

Marcer, P. and W. Schempp (1997) "Model of the Neuron Working by Quantum Holography." *Informatica* 21: 519–534.

Marcer, P. and W. Schempp (1998) "The Brain as a Conscious System." *International Journal of General Systems* 27(1): 231–248.

Marcer, P. and W. Schempp (1999) "Quantum Holography: The Paradigm of Quantum Entanglement," Computing Anticipatory Systems—Second International Conference. D. Dubois (ed.), pp. 461–467. The American Institute of Physics.

Marin, J.M. (2009) "'Mysticism' in Quantum Mechanics: The Forgotten Controversy." *European Journal of Physics* 30: 807–822.

Maruna, S. (2006) *Making Good: How Ex-Convict Reform and Rebuild Their Lives*. Washington, DC: American Psychological Association.

Marx, C. (1992) *Capital*. New York: Penguin Classic.

Massumi, B. (1992) *A User's Guide to Capitalism and Schizophrenia*. London: MIT Press.

Matthew, J.P and M.W. Davidson (2006) http://micro.magnet.fsu.edu/primer/java/interference/doubleslit/.

Matza, D. (1960) *Delinquency and Drift*. New York: Wiley.

Matza, D. (1969) *Becoming Deviant*. Englewood Cliffs, New Jersey: Prentice Hall.

May, T. (1994) *The Political Philosophy of Poststructuralist Anarchism*. University Park, PA: Pennsylvania State University Press.

May, T. (2005) *The Moral Theory of Poststructuralism*. University Park, PA: Pennsylvania State University Press.

McCraty, R. (2002) "Influence of Cardiac Afferent Input on Heart-Brain Synchronization and Cognitive Performance." *International Journal of Psychophysiology* 45(1–2): 72–73.

McCraty, R., M. Atkinson, and R. Bradley (2004a) "Electrophysiological Evidence of Intuition. Part 1. The Surprising Role of the Heart." *The Journal of Alternative and Complementary Medicine* 10(2): 325–336.

McCraty, R., M. Atkinson, and R. Bradley (2004b) "Electrophysiological Evidence of Intuition: Part 2. A System-Wide Process?" *Journal of Alternative and Complementary Medicine* 10(2): 325–336.

McCraty, R., M. Atkinson, D. Tomasino, and R. Bradley (2009) "The Coherent Heart: Heart-Brain Interactions, Psychophysiological Coherence, and the Emergence of System-Wide Order." *Integral Review* 5(2): 10–115.

McTaggart, L. (2008) *The Field: The Quest for the Secret Force of the Universe*. New York: Harper.

Mead, G. (1967) *Mind, Self and Society*. Chicago: University of Chicago Press.

Mender, D. (2010) "Post-Classical Transitions and Emergence in Psychiatry: Beyond George Engel's Model of Psychopathology." *NeuroQuantology* 8(1): 29–36.

Miller, I., and R. Miller (2003) "From Helix to Hologram." http://www.bibliotecapleyades.net/ciencia/ciencia_adn10.htm.

Miller, R., I. Miller and B. Webb. (2002) "Quantum Bioholography: A Review of the Field from 1973–2002. www.emergentmind.org/millerwebbi3a.htm.

Miller, W. (1958) "Lower Class Culture as a Generating Milieu of Gang Delinquency." *Journal of Social Issues* 14: 5–19.

Mills, C.W. (1959). *The Sociological Imagination*. New York: Oxford University Press.

Milovanovic, D. and S. Henry (1991) "Constitutive Penology." *Social Justice* 18: 204–24.

Milovanovic, D. and S. Henry (2001) "Constitutive Definition of Crime," in S. Henry and M. Lanier (eds.) *What is Crime?*, pp. 165–178. Boulder, Colorado: Rowman and Littlefield.

Milovanovic, D. (2002) "Introduction to Pashukanis." In E. Pashukanis, *The General Theory of Law and Marxism*, vi–xxvi, Transaction Books.

Milovanovic, D. (2003) *Sociology of Law*. New York: Criminal Justice Press.

Milovanovic, D. (2004) "Psychoanalytic Semiotics, Chaos, and Rebellious Lawyering." In E. Ragland and D. Milovanovic (Eds.) *Lacan: Topologically Speaking*. New York: Other Press.

Milovanovic, D. (2011) "Justice-Rendering Schemas: A Typology for Forms of Justice and a Prolegomenon for Transformative Justice." *Journal of Theoretical and Philosophical Criminology* 3(1): 1–56.

Milovanovic, D. (2012) "Keynote Address: Constitutive Harm, Justice-Rendering Schemas, and Alternative Construction of Agency (Quantum Consciousness)." Crime and Social Policy in the Caribbean. Derrick Gordon Seminar. University of West Indies at Mona, Jamaica, April 12, 2012.

Milovanovic, D. (2013a) "Quantum Holographic Critical Criminology." *Journal of Theoretical and Philosophical Criminology* 5(2): 1–29.

Milovanovic, D. (2013b) "Postmodernism and Thinking Quantum Holographically." Special Issue, "Crucial Critical Criminology," *Journal of Critical Criminology* 21(3): 341–357.

Mitchell, E. (2008) "Nature's Mind: The Quantum Hologram." Institute of Noetic Sciences. Lake Worth, Florida.

Mitchell, E. and R. Staretz (2011) "The Quantum Hologram and the Nature of Consciousness." *Journal of Cosmology* 14: 1–19.

Mootz, F. (2013) "Is The Rule of Law Possible in a Postmodern World." *Washington Law Review* 68: 249–285.

Murray, Jamie (2013) *Deleuze and Guattari: Emergent Law.* New York: Routledge.

Nakagomi, T. (2003) "Quantum Monadology: A Consistent World Model for Consciousness and Physics." *BioSystems* 69: 27–38.

Nakagomi, T. (2006) "Picture the World as a Quantum Monadistic System." *NeuroQuantology* 4(3): 241–248.

Nelson, R.J. (2012) "Is Virtual Reality Exposure Therapy Effective for Service Members and Veterans Experiencing Combat-Related PTSD?" *Traumatology* 19(3): 171–178.

Nelson, R. J. (2013. "Is Virtual Reality Exposure Therapy Effective for Combat-Related PTSD?" Paper delivered at 2013 Annual Conference, The Society for Social Work and Research. San Diego, CA.

Neuman, Y. (2008) "The Polysemy of the Sign." *Semiotica* 169 (1): 155–168.

Newmann, Von, J. (1996, originally 1932). *The Mathematical Foundations of Quantum Mechanics.* Princeton: Princeton University Press.

Newton, I. (1966, originally 1687). *Principia.* Berkeley: University of California Press.

Nietzsche, F. (2003) *The Genealogy of Morals.* Dover Publications.

Olsen, T., L. Payne, and A. Reiter (2010) *Transitional Justice in Balance: Comparing Processes, Weighing Efficacy.* Washington, D.C.: United States Institute of Peace Press.

O'Malley, P. (2009) *The Currency of Justice.* N.Y. Routledge.

O'Malley, P. (2010) *Crime and Risk.* London: Sage.

O'Malley, P. and S. Mugford (1994) "Crime, Excitement and Modernity." In G. Barak (Ed.) *Varieties of Criminology.* Westport, CT: Praeger.

Parker, C. *Just Lawyers.* Oxford: Oxford University Press.

Pashukanis, E. (2002) *The General Theory of Law and Marxism.* Transaction Books.

Patton, P. (2000) *Deleuze and the Political.* New York: Routledge.

Pavlich, G. (2001) "The Force of Community." In H. Strang and J. Braithwaite (Eds.) *Restorative Justice and Civil Society.* Cambridge; Cambridge University Press.

Pavlich, G. (2005) *Governing Paradoxes of Restorative Justice.* London: Glass House Press.

Pavlich, G. (2007) "Ethics, Universal Principals and Restorative Justice." In G. Johnstone and D.W. Van Ness (Eds.), *Handbook of Restorative Justice.* Cullompton, UK: Willan Publishing.

Penrose, R. (1994) *Shadows of Mind.* Oxford: Oxford University Press.

Penrose, R. and S. Hameroff (2012) "Consciousness in the Universe." *Journal of Cosmology* 14: 1–31.

Pepper, D. (1985) "Non-Linear Optical Phase Conjugation." *Laser handbook 3.* North Holland: Amsterdam.

Perkins, W. (1996) *Droppin Science: Critical Essays on Rap Music and Hip Hop Culture.* Philadelphia: Temple University Press.

Piaget, J. (1995a) "Logical Operations and Social Life." In L. Smith (Ed.) *Sociological Studies*, pp. 134–157. London: Routledge.

Piaget, J. (1995b) "Genetic Logic and Sociology." In L. Smith (Ed.) *Sociological Studies*, pp. 184–214. London: Routledge.

Pierce, J. (1967). *An Introduction to Information Theory.* New York: Dover Press.

Plate, T. (1995) "Holographic Reduced Representations." *IEEE Transactions on Neural Networks* 6: 623–641.

Plate, T. (2003) *Holographic Reduced Representations.* Standford, CA: CSLI Publications.

Poggio, T. (1973) "On Holographic Models of Memory." *Kybernetik* 12: 237–238.

Popp, F. (2000) "Biophotons—Background, Experimental Results, Theoretical Approach and Applications." *Research Advancements in Photochemistry & Photobiology,*" 1: 31–41.

Popp, F. (2012) "Biophotons and Consciousness," Lecture, YouTube: http://meridianenergies.net/tag/alexander-gurwitsch/.

Porter, E. (1991) "The Player and the Dice: Physics and Critical Legal Theory." *Ohio State Law Journal* 52: 1571–1593.

Powell, J. and S. Menendian (2010) "Remaking Law: Moving Beyond Enlightenment Jurisprudence." *Saint Louis University Law Journal* 54: 1035–1,112.

Pribram, K. (1971) *Languages of the Brain.* Monterey, California: Wadsworth.

Pribram, K. (1991) *Brain and Perception: Holonomy and Structure in Figural Processing.* Hilsdale, N.J.: Lawrence and Erlbaum.

Pribram, K. (1999) "Quantum Holography." *Information Sciences* 115: 97–102.

Pribram, K. (2004), http://www.paricenter.com/library/papers/pribram01.php.

Pribram, K. (2006) "Holism vs. Wholism." *World Futures* 62: 42–46.

Pribram, K. (2009) "Physics and Mind." *Zygon* 44(2): 451–466.

Pribram, K. and R. Bradley (1998) "The Brain, the Me, and the I." In M. Ferrari and R.J.Sternberg (Eds.) *Self-Awareness: Its Nature and Development*, pp. 273–307. New York: The Guilford Press.

Prigogine, I. (1981) *From Being to Becoming: Time and Complexity in the Physical Sciences*. San Francisco: W.H. Freeman.

Prigogine, I. (1996) *The End of Certainty: Time, Chaos, and the New Laws of Nature*. New York: The Free Press.

Puthoff, H. (1989a). "Where Does the Zero-Point Energy come From?" *New Scientist* 36; see also: http://newenergytimes.com/v2/archives/fic/N/N19 9402s.PDF.

Puthoff, H. (1989b) "Source of Vacuum Electromagnetic Zero-Point Energy." *Physical Review* 40: 4857–4862.

Quinney, R. (1970) *The Social Reality of Crime*. Little Brown and Company.

Radin, D. (2006) *Entangled Minds: Extrasensory Experiences in a Quantum Reality*. New York: Pocket Books.

Radin, D. (2013) *Supernormal: Science, Yoga, and the Evidence for Extraordinary Psychic Realities*. New York: Deepak Chopra Books.

Ragland, S. and D. Milovanovic (Eds.) (2004) *Lacan: Topologically Speaking*. New York: Other Press.

Rahnama, M., J. Tuszynski, I. Bokkon, M. Cifra, P. Sardar, and V. Salari (2011) "Emission of Mitochondrial Biophotons and Their Effects on Electrical Activity of Membrane Via Microtubules." *Journal of Integrative Neuroscience* 10(1): 65–88.

Rajah, V. (2007) "Resistance as Edgework in Violent Intimate Relationships of Drug Involved Women." *British Journal of Criminology* 47: 196–213.

Rauscher, E. and R. Targ (2001) "The Speed of Thought: Investigation of a Complex Space-Time Metric to Describe Psychic Phenomenon." *Journal of Scientific Exploration* 15(3): 331–354.

Reichenback, H. (1958). *The Philosophy of Space and Time*. N.Y.: Dover Publications.

Robbins, S. (2000) "Bergson, Perception and Gibson." *Journal of Consciousness Studies* 7(5): 23–45.

Robbins, S. (2002). "Sematic Experience and Time." *Cognitive Systems Research* 3(3): 301–337.

Robbins, S. (2004) "On Time, Memory and Dynamic Forms." *Consciousness and Cognition* 13(4): 762–88.

Robbins, S. (2006) "Bergson and the Holographic Theory of Mind." *Phenomenological Cognitive Science* 5: 365–395.

Robbins, S. (2007) "Time, Form and the Limits of Qualia." *Journal of Mind and Behavior* 28(1): 1–22.

Robbins, S. (2010a) "The Case for Qualia." *Journal of Mind and Behavior* 31: 141–156.

Robbins, S. (2010b) "Special Relativity and Perception: The Singular Time of Psychology and Physics." *Journal of Consciousness, Exploration, and Research* 1(5): 500–531.

Robbins, S. (2012) *Time and Memory*. CreateSpace Independent Publishing Platform. https://www.createspace.com/.

Rodowick, D.N. (1997) *Gilles Deleuze's Time Machine*. Durham, NC: Duke University Press.

Rorty, R. (1989) *Contingency, Irony, and Solidarity*. Cambridge, Cambridge University Press.

Rose, T. (1994) *Black Noise: Rap Music and Black Culture in Contemporary America*. Hanover, NH: Wesleyan University Press.

Rosenblum, B. and F. Kuttner (2002). "The Observer In the Quantum Experiment." Foundations of Physics 32(8): 1273–1293.

Rospatt, A. (1995) *The Buddhist Doctrine of Momentariness*. Hamburg: Franz Steiner Verlag Stuttgart.

Rossi-Landi, F. (1977) *Linguistics and Economics*. Amsterdam, Netherlands: Mouton.

Rossi-Landi, F. (1983) *Language as Work and Trade: A Semiotic Homology for Linguistics and Economics*. South Hadley, MA Bergin and Garvey.

Ryle, G. (2002) *The Concept of Mind*. Chicago: University of Chicago Press.

Salmon, W. (1984) *Scientific Explanation and the Causal Structure of the World*. Princeton University Press.

Sampson, R. and J. Laub (1993) *Crime in the Making: Pathways and Turning Points Through Life Course*. Cambridge: Cambridge University Press.

Sartre, J.P. (1972) *Imagination: A Psychological Critique*. Ann Arbor, MI: University of Michigan Press.

Scheff, J. and S. Retzinger (2002) *Emotions and Violence: Shame and Rage in Destructive Conflicts*. London: iUniverse, Inc.

Schempp, W. (1992) "Quantum Holography and Neurocomputer Architecture." *Journal of Mathematical Imaging and Vision* 2: 279–326.

Schempp, W. (1998) *Magnetic Resonance Imaging, Mathematical Foundations and Application*. John Wiley, New York.

Schmidt, C. (2007) *The Concept of the Political*. Chicago: Chicago University Press.

Schroeder, J.L. (2008) *The Four Lacanian Discourses: On Turning Law Inside Out*. New: Routledge.

Schwartz, J., J. Stapp, and M. Beauregard (2004). "Quantum Physics in Neuroscience and Psychology." *Philosophical Transactions*, www.physics.lbl.gov/~stapp/ptrs.pdg.

Scully, M.D. and K. Druhl (1982) "Quantum Erasure: A Proposed Photon Correlation Experiment Concerning Observation and 'Delayed Choice' in Quantum Mechanics." *Physical Review* 25: 2208–2213.

Searle, J. (1970) *Speech Act Theory: An Essay in Philosophy of Language*. Cambridge University Press.

Shannon, C. (1948) "A Mathematical Theory of Communication." *Bell System Technical Journal* 27: 379–423, 623–656. Available in pdf: http://cm.bell-labs.com/cm/ms/what/shannonday/shannon1948.pdf.

Shannon, C. and W. Weaver (1998) *The Mathematical Theory of Communication*. Urbana, IL: University of Illinois Press.

Shaw, C. and H. McKay (1969) *Juvenile Delinquency and Urban Areas*. Chicago: University of Chicago Press.

Sheehan, D. (ed.) (2006) *Frontiers of Time: Retrocausation—Experimental and Theory*. AIP Conference Proceedings. Melville: New York.

Shimony, A. (1965) "Quantum Physics and the Philosophy of Whitehead," in M. Black (ed.), *Philosophy in America*, pp. 240–261. New York: Cornell University Press.

Sippel, L. (2013) "How Does Virtual-Reality Therapy for PTSD Work?" *Scientific American* 24(5): http://www.scientificamerican.com/article/how-does-virtual-reality-therapy-fo/.

Sklar, L (1992) *Philosophy of Physics*. Westview Press.

Smolin, L. (2001) *Three Roads to Quantum Gravity*. New York: Basic Books.

Smolin, L. (2013) *Time Reborn: From the Crisis in Physics to the Future of the Universe*. New York: Houghton Mifflin Harcourt.

Soja, E. (2011) *Postmodern Geographies: The Reassertion of Space in Critical Social Theory*. London: Verso.

Spinoza, B. (1985) *Complete Works*. E. Curley (Ed. and Transl.). Princeton: Princeton University Press.

Stapp, H. (2007. *Mindful Universe: Quantum Mechanics and the Participating Observer*. Springer-Verlag.

Stapp, H. (2008) "Relativistic Whiteheadian Quantum Field Theory," http://www-physics.lbl.gov/~stapp/stappfiles.html.

Stapp, H. (2010) "Quantum Mechanical Approach to the Connection Between Mind and Brain." Listed in: http://www-physics.lbl.gov/~stapp/stappfiles.html.

Stapp, H. (2012) "Reply to a Critic: 'Mind Efforts, Quantum Zeno Effect and Environmental Decoherence.'" *NeuroQuantology* 10(4): 601–605.

Stapp, H. (2013a) "Quantum Theory of Consciousness." Paris Talks (see his homepage).

Stapp, H. (2013b) "Quantum Physics and Philosophy of Mind." Milan Lecture. www-physics.ibl.gov/~stapp/stappfiles.html.

Suarez, A. (2007) "Classical Demons and Quantum Angeles: On 't Hooft's Deterministic Quantum Mechanics." arXiv:0705.3974v4.

Sullivan, D. and L. Tifft (2005) *Restorative Justice: Healing the Foundations of Our Everyday Lives.* Monsey, New York: Criminal Justice Press.

Susskind, L. (1995) "The World as a Hologram." *Journal of Mathematical Physics* 36: arXiv:hep-th/9409089.

Susskind, L. (2003) "Black Holes and the Information Paradox." *Scientific American*, Special Edition, March: 18–23.

Susskind, L. (2009). *The Black Hole War.* New York: Little, Brown and Company.

Susskind, L. and J. Lindesay ((2011) *An Introduction to Black Holes, information and the String Theory Revolution.* London: World Scientific.

Susskind, L. and E. Witten (n.d.) "The Holographic Bound in Anti-de Sitter Space." hep-th/d9805114.

Sutherland, E. (1974) *Criminology.* J.B. Lippincott Company.

Talbot, M. (1991) *The Holographic Universe.* New York: Harper Collins.

Thornberry, T. (1987) *"Toward and Interactional Theory of Delinquency." Criminology* 25: 863–891.

Thornberry, T. and M. Krohn (2005) "Applying Interactional Theory to the Explanation of Continuity and Change in AntiSocial Behavior," in D. Farrington (Ed.) *Integrated Developmental and Life Course Theories of Offending*, pp. 183–201. New Brunswick, NY: Transaction Books.

Thorne, K.S., R.H. Price, and D.M. MacDonald (1986) *Black Holes, The Membrane Paradigm.* New Haven: Yale University Press.

't Hooft, G. (1993) "Dimensional Reduction in Quantum Gravity." arXiv:gr-qc/9310026.

't Hooft, G. (2007) "On the Free-Will Postulate in Quantum Mechanics." arXiv:quant-ph/0701097v1.

Tiller, W., R. McCraty, and M. Atkinson (1996) "Cardiac Coherence." *Alternative Therapies in Health and Medicine* 2(1): 52–65.

Unger, R. *What Should Legal Analysis Become?* New York: Verso.

Unger, R. (2004) *False Necessity.* N.Y. Verso.

Verlinde, E. (2010) "On the Origin of Gravity and the Laws of Newton" arXiv:1001.0785v1.

Virtual Reality Medical Center (n.d.) "About Virtual Reality." www.rphobia.com/about.htm.

Von Newmann, J. (1996) *Mathematical Foundations of Quantum Physics.* Princeton University Press.

Wadsworth, T. (2004) "Industrial Composition, Labor Markets, and Crime." *Sociological Focus* 37: 1–24.

Walborn, S.P., M.O. Cunha, S. Padua, and C.H. Monken (2002) "Double-Slit Quantum Erasure." *Physical Review*: 65: 033818-1–033818-6.

Wallack, F. (1980) *The Epochal Nature of Process in Whiteheads Metaphysics.* Albany, New York: State University of New York Press.

Ward, T. (2010) "Commentary: Disciplinary Segregation in Prisoners Dignity, Virtue, and Punishment: The Ethical Justification of Disciplinary Segregation in Prisoners." *Journal of Theoretical and Philosophical Criminology* 2(3): 98–109.

Watkins, S. C. (2001) "A Nation of Millions: Hip Hop Culture and the Legacy of Black Nationalism." *The Communication Review* 4: 373–398.

Weber, M. (1978) *Economy and Society.* Vols. 1–2. G. Roth and C. Wittich (Eds.), Los Angeles: University of California Press.

Wegner, D. (2002) *The Illusion of Conscious Will.* Cambridge, MA: MIT Press.

Weisburd, D, and A. Piquero (2008) "How Well do Criminologists Explain Crime?" in M. Tonry (ed.) *Crime and Justice: A Review of Research*, pp. 153–502. Chicago University Press.

Wendt, A. (2006) "An Auto-critique from a Quantum Perspective," in S. Guzzini and A. Leander (eds.) *Constructivism and International Relations*, pp. 181–219. London: Routledge.

Wendt, A. (2010) "Flatland: Quantum Mind and the International System of Hologram," in M. Albert, L. Cederman and A. Wendt (eds.) *New Systems Theories of World Politics*, pp. 279–310. London: Palgrave.

Wheeler, A.J. (1978) "The 'Past' and the 'Delayed-Choice Double-Slit Experiment'," in A.R. Marlow (Ed.), *Mathematical Foundations of Quantum Theory*, pp. 9–48. New York: Academic Press.

Wheeler, J. (1983) "Delayed Choice." in J.A. Wheeler and W.H. Zurek (Eds.), *Quantum Theory and Measurement.* Princeton: Princeton University Press.

Wheeler, J. (1990) "Information, Physics, Quantum: The Search for Links," in W. Zurek (ed.) *Complexity, Entropy, and the Physics of Information*, pp. 3–28. Addison-Wesley.

Whitehead, A. (1925) *Science and the Modern World.* New York: Mentor Books.

Whitehead, A. (1929) *Process and Reality.* Cambridge: Cambridge University Press.

Whorf, B. (1964) *Language, Thought and Reality.* MIT Press.

Widdows, D. (2004) *Geometry and Meaning.* Stanford, CA: CSLI Publications.

Wieder, L.(1988) *Language and Social Reality.* Landham, MD: University Press of America.

Wilson, W. (1996) *When Work Disappears: The World of the New Urban Poor*. New York: Vintage.

Wilson, S. (2013) "Veterans and the Digital Battlefield." *DAV Magazine*, November/December: 16–18.

Witten, E. (1998) "Anti De Sitter Space and Holography." *Theoretical Mathematical Physics* 2: 253–291.

Wittgenstein, L. (2009) *Philosophical Investigations*. West Sussex, UK: John Wiley and Sons.

Woolford, A. (2009) *The Politics of Restorative Justice*. Winnepeg, Canada: Fernwood Publishing.

Woolford, A. and R.S. Rattner (2008) *Informal Reckoning: Conflict Resolution in Mediation, Restorative Justice and Reparation*. London: Routledge-Cavendish.

Yager, M. (2013) "E-Discover as Quantum law: Clash of Cultures—What the Future Portends." *Richmond Journal of Law and Technology* 19: 10–22.

Young, J. (2011) *The Criminological Imagination*. New York: Polity Press.

Young, T. (1804) "Experimental Demonstration of the General Law of the Interference of Light." *Philosophical Transactions of the Royal Society of London*, vol. 94 (1804).

Index